WE LIVE IN THE ORBIT OF BEINGS GREATER THAN US

Patrick Farnsworth

Interviews from Last Born in the Wilderness

WE LIVE IN THE ORBIT OF BEINGS GREATER THAN US

ISBN 978-1-7325523-7-1

First Printed
by Gods&Radicals Press 2020

Cover Design & Layout by Casandra Johns: houseofhands.net
Editing Team Rhyd Wildermuth, Dani'l Nighting
Cover Image: CC0 Mark Basarab on Unsplash.
Interior Images: CC0 Adrien Olichon, Aleksey Sokolenko, Anna Popovic, Annie Spratt, Aperture Vintage, Bianca Castillo, Casey Horner Claudio Testa, Claus Grunstaudl, Curioso Photography, Daniele Levis Pelusi, David Clode, Eberhard Grossgasteiger, Eriks Abzinovs, Georgia de Lotz, Jo Wroten, Julien Leclercq, Justin Kauffman, Kelly Sikkema, Khai Truong, Kirill Pershin, Laura Vinck, Meagan H, Michael Olsen, Mike Ko, Milada Vigerova, Nathalia Rosa, Nathan Dumlao, Patrick Fore, Paul Talbot, Pawel Czerwinski, Ramin Khatibi, Scott Webb, Sina Katirachi, Spencer Watson, Tamas Tuzes Katai, Tim Mossholder, Velimir Matic, and Yogesh Pedamkar on Unsplash.

Gods&Radical Press
PO Box 11850
Olympia Washington 98508

Solidarity, bulk discount, and wholesale copies available. Contact the editors or author, or for distro: distro@abeautifulresistance.com

Gods&Radicals Press is a not-for-profit anti-capitalist Pagan publisher. View our other works and online journal at abeautifulresistance.org

CONTENTS

"HEY PATRICK, IT'S JOHN HALSTEAD IN NORTHWEST INDIANA.

"JUST WANTED TO CALL AND THANK YOU FOR THE WORK THAT YOU'RE DOING. ESPECIALLY, I LISTENED TO THE 200TH EPISODE, WHICH WAS THE COLLECTION OF DIFFERENT INTERVIEWS. I THOUGHT IT WAS REALLY, REALLY AMAZING HOW YOU MADE THESE SEEMINGLY DISPARATE IDEAS JUST BUILD ON EACH OTHER. I LOVED THAT YOU ENDED IT WITH BAYO AKOMOLAFE BECAUSE HE'S JUST AMAZING...

"I JUST WANTED TO MENTION I THOUGHT IT WAS SO COOL, AND I WAS THINKING THIS THERE'S A BOOK HERE, YOU KNOW. THERE'S A COLLECTION OF ESSAYS THAT YOU COULD PUBLISH. I JUST HOPE YOU CONSIDER DOING IT...

"THANKS FOR THE GREAT WORK YOU'RE DOING. I LOVE YOUR PODCAST AND I HOPE YOU KEEP IT UP. I'M GLAD YOU CAME BACK FROM YOUR VACATION. THANKS A LOT, BYE."

JOHN HALSTEAD, 8:53 AM MST, AUGUST 18TH 2019

INTRODUCTION
WILDERNESS OF MINE AFFLICTIONS

WITH THIS VERY GENTLE SUGGESTION, the seeds of this book were planted. Or maybe the better way to say it is that the seeds were planted long ago, and John's insightful, passing suggestion was the little charge of life that was needed for this idea to emerge, become fully realized, and to be ready to live on its own.

As John mentions in that call, this book is largely based on the 200th episode of my podcast *Last Born in the Wilderness*, a project I've been engrossed in producing for the better part of a decade. The episode John mentioned is a nearly four-hour-long audio essay, in which I attempted to present my vision of what this ongoing project has been for me (and I assume for others, as well), weaving together about nineteen segments (twenty if you count my TEDx talk, *Forging Connection in Perilous Times*) of interviews I had conducted over the past one hundred episodes.

Each segment was tied together with commentary—my attempt to connect the various threads of insight from the wide-ranging voices I've presented in this work, binding seemingly disparate topics and perspectives together into a coherent vision.

As the host of this project, I conduct long-form interviews; I then individually record, edit, and release each as a weekly stand-alone episode. I find people I'm particularly interested in, whom I feel would serve the spirit of this work, and ask them to participate by presenting their ideas. While I may highlight very different perspectives through this process, I've always felt there's a common thread that runs through each of them, something I attempted to demonstrate in the release of that 200th episode.

To say that this is a work of journalism, or that I'm presenting some kind of objective stance on what's happening in the world would, I think, be arrogant. It would also miss the point entirely. Certainly, there are scientific facts and rigorous forms of research presented in many of these interviews: in particular, regarding abrupt climate disruption and the unraveling of the biosphere (what I've occasionally and blandly referred to as "environmental issues") as a direct result of unrestrained industrialization over the past several centuries. These discussions also explore what's happening politically, socially, and culturally around the world as these trends unfold. Those topics are discussed with a certain amount of factual basis.

But I want to make it absolutely clear that I have an agenda. I am biased in my presentation of these subjects, and the aims of this project should be apparent to anyone that has come across my work and delved into the catalogue of interviews I've conducted.

In order to further my vision with this work, I think it might be necessary to explain the title of the podcast in some detail. There are a few different layers to it, and I think this will convey my intentions and what this book in particular is meant to present.

First, the title *Last Born in the Wilderness* is a direct reference to a passage in the Book of Mormon (yes, that's right). To directly quote the passage:

> And now I speak unto you, Joseph, my last-born. Thou wast born in the wilderness of mine afflictions; yea, in the days of my greatest sorrow did thy mother bear thee. (2 Nephi, 3:1)

I grew up as a member of The Church of Jesus Christ of Latter Day Saints (LDS or the Mormon church, informally), embedded deeply within its insular community and culture in Southern Idaho. The reason, at least on the surface, that I'm alive and living on this land we now call the United States of America is because my European ancestors converted to the gospel of the LDS Church sometime in the mid-to-late nineteenth century. They then sailed across the Atlantic Ocean, and through what I imagine was an arduous journey, migrated across the great and expansive North American continent. They then settled in the region now formally defined as the state of Utah, a region then under the governorship of Brigham Young, the second leader, or "prophet, seer, and revelator," of the LDS church.

I left the church when I was about seventeen or eighteen years old. I came to my senses, so to speak. I was the youngest of six children, and my father, quirky religious man that he is, would often refer to me as "my last born in the wilderness," in direct reference to that passage in the Book of Mormon I quoted above.

What I didn't anticipate is that, as the themes of the podcast evolved over the years (and quite some time after shaking off my membership in the church), I began to realize that the subjects I was interested in talking about and exploring with others were really apocalyptic in nature (remember, the root of apocalypse, the Greek word *apokálypsis,* means "uncovering, disclosure, revelation"). Apocalypse especially fits when it comes to the subject of abrupt climate disruption and what this trend has presented for the future of our species and for nonhuman life on this planet. The systems required to maintain the habitability of this planet are being disrupted and altered very rapidly and dramatically, with dire implications. And so within that context, what does it mean to be the "last born in the wilderness"?

For instance, is it the wilderness that is disappearing, is it us, or both? Are we really the last people on the planet to actually experience these wild and intact places, to have an actual conception of the wilderness and an experience of it? In actuality, yes: these places are disappearing at an alarming rate, with very little to indicate that this will be, or can be, reversed.

That's a very serious proposition that I pose with this work. It is within this context and within this space that I present these interviews, regardless of the person that I'm interviewing and their particular perspective on that subject.

And truthfully, it's a difficult thing to grapple with. Obviously, you can go over the information and figure it out for yourself. There are mountains of data that point to that conclusion, which we will get into in the first section of this book. But to understand it on an intellectual level is really just the first step in the process of radically accepting this information.

It's one thing to come to that conclusion objectively and to see the information as it is, and it's another thing entirely to accept it on an emotional, spiritual, and animal level, to really know what it means to be alive in a time like this.

I come to this work with a sense of empathy and compassion for those I am living with on this planet, for my fellow brothers and sisters that were born in this time. When I first started this work, I felt I had to present this information aggressively, that I would have to shout this truth from the mountaintop, to say unabashedly: "We're going to go extinct, so straighten the fuck up and act accordingly (whatever that means)."

But I realized that just as much as this is a time to speak the truth, this is also a time to be very gentle and humble, to understand that people are now seeing the writing on the wall in one way or the other—whether they consciously acknowledge it or not. It's very important to understand the feelings that come up when the truth of our collective predicament is faced, and that our reactions can lead us in very different directions.

How you, how we, choose to proceed with this knowledge is the most important question for us right now.

This project for me has now become, more than anything, a space that I and many others have co-creatively carved out for those ready to have this discussion, those ready to talk about what's really happening, and to talk about how to be in this time.

So much of what we're hearing right now in climate and environmental activism is framed within the imperative to act. Action, action, action! We have to do something right now, now, now! As it is often framed, we only have twelve, or ten, or five years left before the climate system hits some arbitrary, politically agreeable threshold of no return, a Rubicon that once crossed, we'll never be able to come back from.

The truth is, we've already crossed the Rubicon. Regardless of what we do right now, global climate change is happening, with enough heat baked into the system to guarantee several degrees of warming just from the heat trapped in the oceans alone. That heat is irrevocable, and the level and rate of change currently underway is likely beyond our capacity to survive.

I'm not saying that people shouldn't do something, that there aren't actions that are meaningful, important, or necessary right now. The point here is to ask the question: what kind of actions are we supposed to engage in now, within the proper context and understanding that I and so many others have outlined?

If you want to engage in political action and social change, what kind of action is worthy of our love, dedication, and rage in a time like this? Whatever it may be, how you treat your fellow human beings and nonhuman life on this planet has to be within the understanding that we may not have much time left. And so how do you want to be? What is the most sacred and beautiful way that you can be in this time, with the full awareness that we don't really have a lot of time to fuck around?

This is by no means meant to disempower you. There is a delusional view going around that hope itself is a facilitator for action. In my view, hope—especially hope grounded in the logic of this system—occludes us from engaging in the actions that are required of us right now. As my friend Barbara Cecil has said, "Hope is fundamentally an avoidance of an essential risk that needs to be taken."

Imagine you were given a terminal diagnosis for some disease, and the doctor comes into your hospital room and begins to tell you that this is the time to try a series of intensive surgeries and treatments. The doctor says that by doing this, you will survive and beat this thing. But you know, deep down, that this isn't true. Not only that, everyone close to you can see what's really happening as well.

In these circumstances, what is the more honorable and right thing to do: exist in a state of denial about the finite quality of your life and believe that you can "beat" this thing? Or instead, accept that the time you have left is precious, that it is imperative to embody the knowledge of your death, be present in that fact, and accept that grief itself is a facilitator for right action, whatever that may be?

When I talk about action, I'm coming from a radical perspective. When I interview people about what we can do, I'm not looking at it from this sort of institutionalized perspective that we often hear in mainstream activism, this nonviolent, defanged thing where we appeal to authorities and governments to do the work for us. They're not going to do it for us. It doesn't actually benefit them to do it, whatever "it" may be.

Whatever proposals, agreements, or legislation they present will never get to the root of the matter. They want to keep the system going as long as possible, even if that means further plundering the earth, exploiting and extracting as much as they possibly can before the big game is up. They're going to continue to do that unabated, except they are going to call it "green," "sustainable," or even "compassionate" and "conscious," when it's really just more of the same. As my friend Dahr Jamail would say it, it's just "capitalism with a green leaf on it."

The kind of actions I suggest we engage in may seem a bit radical and extreme, but considering where we're at right now, that's really the only meaningful type of action that I can see worth engaging in. I really just want to provide a space for people to comprehend, understand, and learn to be present, to not be in denial, to recognize that it is your sacred duty right now to be as present as possible, and every action that you take must come from that place of knowing.

As they say, once you see it, you can't unsee it.

It isn't a mistake that you're here right now. It isn't a mistake that I'm here right now. We're supposed to be here right now. We must take responsibility for where we are, and part of taking responsibility is to understand what's actually required of us.

In doing this work, I've tried to contribute in some way to that discussion. I hope in exploring this work, you too can begin to figure those things for yourself. You don't have to agree with me or with anybody I've presented in this book. I only hope that the perspectives that I highlight will help you come to terms with what's currently unfolding on this planet, and that this is no time to look away.

PART ONE
DEATH BY A THOUSAND CUTS

A PIVOTAL MOMENT in my journey was in April 2018, when I gave my first public talk. It was the first time I presented my ideas and work publicly.

I was really saying it out loud, on a stage, that we are in the midst of a mass extinction event, which likely includes us in that process.

The co-organizers of the TEDxTwinFalls event Way, Shape, Form, Mykelle Walton and Jennifer Thornquest, encouraged me to speak my truth, the truth, bluntly. Jennifer, as my speaking coach, gave me the permission to go there, for which I'm deeply grateful.

The title of that talk was *Forging Connections in Perilous Times*, and it was presented and recorded at the Orpheum Theater in Twin Falls, Idaho. This is an excerpt:

> A topic that I discuss quite regularly and frequently on this podcast can be called "environmental issues." I think that is a rather neutered term. I like to more accurately describe it as the systematic unraveling of our biosphere or the life systems of our planet. I think it's not quite as benign as "environmental issues," but I do think it is much more accurate as a descriptor.
>
> As it stands today, our culture is a life-destroying force on this planet. Since 1970, global wildlife populations have declined by over 50 percent. Insect populations worldwide have seen massive declines these past several decades as well. Our oceans are in dire straits: the excessive amounts of carbon that have been produced by human industrial activity are affecting them, making them more acidic and essentially inhospitable for life to exist. And that doesn't even include the billions of tons of plastics and toxic waste that we've been dumping into our oceans. In the global north the permafrost, which is of course the frozen soil, or I should say, previously frozen soil, is thawing. As it thaws it is releasing massive quantities of carbon and methane into our atmosphere, further exacerbating global climate change, which is again a result of human industrial activity these past few centuries.

DAHR JAMAIL

I would like to present a segment of my second interview with Dahr Jamail. Dahr is an award winning journalist who, until very recently, reported regularly on climate disruption and environmental issues for the online publication Truthout, and is the author of *The End of Ice: Bearing Witness and Finding Meaning in the Path of Climate Disruption,* released January 2019.

In October 2018, I did a live podcast event in which people could see the event in person as well as through a live-stream. Two friends, Jordan Thornquest and Chase Chandler, helped film and coordinate the event, and because of their expertise and dedication, the whole thing went off without a hitch.

I asked Dahr to participate in the event so he could help explain, through his years of in-depth research and reporting on abrupt climate disruption, what is currently unfolding in our global climate system. In that previous segment of my TEDx talk, I briefly explained a few of the major trends, but it's hard, within the limited timeframe of a nine-minute talk, to unpack all that information and get to the major points I was trying to present. This live event with Dahr was an attempt to delve more deeply into this subject and to examine these trends more closely.

In this segment of my live interview with Dahr Jamail, we begin by discussing the IPCC report released in 2018.

PATRICK FARNSWORTH: IPCC (which stands for the Intergovernmental Panel on Climate Change) is a UN body of scientists that releases a report every few years. They just released a report within the last month or so, and I want to quote from the press release. This is discussing the differences between a 1.5 centigrade to a 2 degree centigrade global average temperature rise. So, from the press release:

> The report highlights a number of climate change impacts that could be avoided by limiting global warming to 1.5°C compared to 2°C or more. For instance, by 2100, global sea level rise would be 10 centimeters lower with

> global warming of 1.5°C compared with 2°C. The likelihood of an Arctic Ocean free of sea ice in summer would be once per century with global warming of 1.5°C, compared with at least once per decade with 2°C. Coral reefs would decline by 70 to 90 percent with global warming of 1.5°C, whereas virtually all ninety-nine percent would be lost with a 2°C rise in temperature. "Every extra bit of warming matters, especially since warming of 1.5°C or higher increases the risk associated with long-lasting or irreversible changes such as the loss of some ecosystems," said Hans-Otto Pörtner, the co-chair of IPCC Working Group 2.

So the thing about the IPCC: this is actually a conservative report, yet even for as conservative as it is, it's still quite dire. What is your opinion on that report? What do you make of the comparisons between 1.5 to 2°C in temperature?

DAHR JAMAIL: I would echo what you just said.

The thing about this report that I think people need to understand is that, as alarming and dire as it is, there's nothing new in it. They didn't generate any new information. They basically just took what they already had and put it together to try to resound the alarm they had hoped previous reports had sounded, which is the absolute necessity to limit warming to 1.5°C as opposed to 2°C or 4°C (which is the trajectory we're currently on), because there is a massive difference in global implications between 1.5°C warming and 2°C warming—as you just read.

So as dire as this is, I just want to reiterate this is basically science by committee, if you will, in the IPCC. This is very old data. Some of what they're using to make these projections and assessments is more than 10 years old. And then when they take all that data into their report, it's an extremely politicized body. Then, you have a country like the United States coming in and, of course, lobbying with everything they're worth to go with the least severe ramifications of this data, instead of really showing what could happen and warning against the worst case scenario.

This is why, with every IPCC assessment that's come out, reality has actually outpaced its worst case, future projections. As we catch up with what they had projected for the year 2000 or the year 2010 or coming up to 2020, we're actually ahead of all their previous worst case projections. That's a trend that I fully expect to continue into the future. And not just me, but some of the authors of IPCC reports whom I've interviewed as well.

PATRICK FARNSWORTH: Right. And something they mentioned in the part that I quoted was an ice-free Arctic. From what I can tell, just based on the graphs that come out every year, that sea ice extent just diminishes pretty dramatically. Could you please explain the significance?

DAHR JAMAIL: Well, the current projection is that we're set up to start having ice-free periods in the summer of the Arctic within about five years, and it could be sooner. The current projections looking at trends and minimum extents and minimum volumes, etc., show it to be beginning to happen within five years from now. Why so many people are fixated on Arctic sea ice—myself included—is because there's several very worrisome things that are set to happen once that's gone. We've already actually started seeing them happen, but they will only accelerate when there are periods of no Arctic sea ice in the summer.

The first and most important is that sea ice is basically reflecting back sunlight and the heat sunlight creates with it. This keeps the heat from going down into the water, getting into the shallow seas, and warming the massive amounts of methane hydrates that are frozen into that sub-sea permafrost. Once that ice starts melting, the methane released is 100 times more potent as a greenhouse gas than carbon dioxide on a 10-year timescale. Methane remains 20 times as potent even on a 100-year timescale.

Once that methane starts getting released en masse, we're essentially replicating the key extinction driver during the Permian mass extinction 252 million years ago, which wiped out 90 percent of all life on Earth—most of it in the oceans. We are on track to replicating that even this summer. We've seen lakes now in the Arctic that are literally bubbling because the permafrost underneath them is melting and releasing methane that's coming up through the water. Other sub-sea areas are also already seeing releases.

This has been happening for several years now, but it's now accelerating. For this to happen across the Arctic, especially once we have that sea ice completely gone for periods during the summer, means we're at grave risk of that happening all across the shallow seas of the Arctic.

So that's number one. Number two: Arctic ice melt shifts weather patterns globally. It's already causing Greenland melts, and that steady trend of overall melting in Greenland is continuing. Of course, it vacillates year to year, but overall we see a consistent downward ice trend and increasing melting in Greenland. That's causing a disruption of the Atlantic Meridional Ocean Circulation current, the AMOC, which is a giant circulating current from the Atlantic. As that slows down, it will affect weather patterns dramatically, particularly in northwestern and northern Europe.

So there are to be major weather pattern shifts. Along with that comes rainfall patterns shifting across the globe. This means food supplies are going to be impacted. That's all still a very cursory glance, but it gives you an idea of how important it is that this sea ice remains intact. The unfortunate reality we're looking at is that the things I described appear to be inevitable. I think it's just a question of when.

THE UNFORTUNATE REALITY WE'RE LOOKING AT IS THAT THE THINGS I DESCRIBED APPEAR TO BE INEVITABLE. I THINK IT'S JUST A QUESTION OF WHEN.

Everything that Dahr said is coming to pass. As I stated previously, that interview was recorded in October 2018. If you look at any of the graphs and data that have come out just this past year alone, you would see a huge spike of ice loss and temperature rise, specifically in the Arctic region. It's incredibly disturbing to see what's currently unfolding on this planet right now, how rapidly global weather patterns are shifting and how ocean currents are slowing down, and what all this implies for our species as we proceed into the future.

The direct implication of the information Dahr has presented in his work is that the human species will be unable to adapt to the rapid shifts from abrupt climate disruption. This is in great part due to a growing inability to grow food at scale, along with the widespread conflict that will ensue as a result of resource depletion, catastrophic weather change, dramatic sea-level rise inundating coastal cities, large displacement of human populations, and other widespread and unanticipated changes relating to abrupt climate disruption and ecological collapse.

NICHOLAS HUMPHREY

I'd like to further this point by discussing how these shifts are already impacting food production by presenting a segment of my interview with interdisciplinary scientist Nicholas Humphrey. Nicholas Humphrey is a meteorologist and geoscientist, with a focus on extreme weather events and their connection to our destabilizing climate. Nick's goal is to communicate, in an interdisciplinary way, the serious risks associated with climate tipping points, extreme weather events, and ecological collapse.

When I spoke with him in April 2019, we discussed the record-breaking floods in the Midwestern United States that occurred only weeks prior to the recording of that interview. To give some background to this event: record-breaking snowfall in the Midwest, with severe fluctuations in temperature in the late winter and early spring of 2019, coincided with a massive "bomb cyclone" that visited the region, leading to a rapid and massive overflow of the Missouri River. This inundated numerous farms, cities, and towns and overwhelmed much of the overall infrastructure to the point of collapse. Many farmers in the region still have not recovered and will take years to do so.

But as Nicholas explains, the frequency of these events has a compounding effect. It's one thing for these kinds of events to happen every decade or so, but it's another thing entirely for them to happen more frequently and severely as time goes on.

NICHOLAS HUMPHREY: My actual bachelor degree is from South Dakota State. It's in interdisciplinary studies, so I learned theoretical research concepts and how to take a complex, multicausal problem and look at it through the eyes of different disciplines that tackle it. Then, I try to understand it from these individual perspectives and fuse those perspectives together into a new concept, a new idea. This is what some refer to as a complex and multicausal explanation, or basically an interdisciplinary model or theory of how things are functioning.

Climate change is really one of those things for which the causative agent is very clear. It's something that can be discussed from an atmospheric chemistry perspective, but all the effects from all of the feedbacks, and everything that encompasses abrupt climate change or disruption, and the damage of it all: that's really a multidisciplinary thing.

I take the approach of trying to make sure people understand just how damaging these events are, and the acceleration of these events, and why it's accelerating.

I saw something written by the World Economic Forum about how insect pollinators contribute 153 billion dollars a year to the global economy through crop production. But the diversity and populations of insects are decreasing, threatening crop production. And I was like, well: so okay, you lose crops but what happens to the people that need the food? What happens to the people that produce the food? What happens to the ability to even

CLIMATE CHANGE IS REALLY ONE OF THOSE THINGS FOR WHICH THE CAUSATIVE AGENT IS VERY CLEAR.

have current agriculture when you have both pollinators decreasing and storms getting worse, destroying areas that you needed to plant your food?

There's always this economic numbers game, or it's the temperature changes, or it's the soil. I'm not an expert in everything, but I can take what these experts are saying and try to put it together into something more comprehensive to make people understand how significant these things are, because that's really not being done.

As a result of that not being done, the average person really doesn't get how bad it is, how bad it's becoming, and how fast it's all happening. Right now I'm watching the Arctic undergo what is basically a nonending heatwave. We can't even call it a heat wave anymore: it's just gotten very warm since about late February or early March, and it's not cooling down, and it's catastrophically losing sea ice. The permafrost is melting exponentially, meaning it's just getting worse and worse every year.

Nobody really talks about this. No one knows this is happening, because nobody tells them. So nobody really understands the implications of this, or the fact that climate change is going to continue accelerating significantly in the coming years and we're gonna have to deal with more of these events. Imagine events like the Midwest flooding happening every couple of years. I mean: happening every couple of years, not even every year or more than once a year—just every two to five years.

We already had the Missouri flood in 2011. It took forever for many places to recover, and now we have this one in 2019. Imagine even those events happening every couple of years. Or you go from this to a huge drought that doesn't end for four or five years, and then we get hit with another major flood. People don't understand that it's about an accumulation of effects, and so most people just go walking around like everything's fine. "Oh, is this a bad flood." "Oh you know this has happened before." It's a lot of normalization.

I don't have very much optimism for the next four or five years in terms of what's going to happen to this region. I mean, I'm worried about this summer. I'm worried about having this flood and then having severe storms. Even if the storms themselves aren't particularly notable, we will get more rain and more flooding events that exacerbate everything that's already happened. Then maybe in the summer it just gets really hot, and so the crops we're still able to grow have to deal with heat stress and drought.

It gets harder to predict beyond the immediate future what's going to happen: when the climate is changing so rapidly, you can't really depend on traditional ways of projecting the future reliably anymore. That's the biggest danger. People aren't going to really know what's going to happen next and be able to prepare for it.

As Dahr and Nicholas lay out in their respective interviews, abrupt climate disruption is an obvious threat to the continuity of our way of living on this planet—ecologically, economically, and materially. The mass production of food, generated and maintained by international distribution networks, is already being severely disrupted by the impacts of anthropogenic climate disruption, and disruption will only increase as these trends accelerate.

FRANCISCO SÁNCHEZ-BAYO

As true as this is, there are other factors to consider within this trend of ecological disintegration we are witnessing, in particular the impact that expansion of urban areas and industrial agricultural projects is having on communities of life, including insects.

For more on this point, here is a segment of my interview with environmental scientist and ecologist Francisco Sánchez-Bayo. Francisco's extensive research has revealed the drivers that are causing the severe loss of biodiversity for insects around the world, directly related to chemically intensive monocultural agricultural practices, as examined in his scientific survey "Worldwide Decline of the Entomofauna: A Review of Its Drivers," coauthored with Kris Wyckhuys.

That survey reviews dozens of contemporary studies from around the world, each which document the rapid decline of global insect populations in their respective localities. The studies indicate that most species of insects (in particular pollinators, which are fundamental for the survival of crops and plant life in general) are likely to go extinct by the end of the century.

Deforestation and the intensive use of insecticides, herbicides, and chemical fertilizers are the main drivers behind this. As agricultural land conversion continues unabated around the world, especially in the most biodiverse regions on the planet, we will continue to see wide-ranging damage to ecosystems, with 41 percent of insects having disappeared over just the past several decades.

FRANCISCO SÁNCHEZ-BAYO: As we point out in our review, apart from pointing to the numbers and the species that are declining, some of them are increasing as a result of the climate. We also analyzed their drivers as indicated by each one of the studies, and also our personal observations, and we back up all those drivers with other information that is available in the literature.

The main driver is basically the loss of habitat—many habitats, many ecosystems—that we are destroying, starting with the tropical forest.

In the tropical areas there is the draining of wetlands in many developing countries and the expansion of urban areas and industries which have taken over more and more land. That loss of habitat obviously is the main driver.

One of the major factors of this destruction is really agriculture. The reason why many tropical forests are cut down is simply because of agriculture, and not just to get some timber for exporting to developed countries. It's to get more land to grow soybeans or other crops, as in Brazil, for example. We know that.

And in Southeast Asia, the destruction of tropical forests is just rampant just to plant crops. Agriculture is the driver behind that. With agriculture comes the use of pesticides and fertilizers. Pesticides obviously are insecticides, and insecticides are chemicals designed to kill insects. No one can deny that pesticides—in particular insecticides—are having a huge impact and are one of the drivers for the decline of insects. It is obvious.

Then there are other chemicals which are toxic to insects: industrial chemicals and discharges from mines and industries that contaminate the waterways. So all the insects living in rivers and ponds and lakes are affected by these chemicals. But they are not as much affected as they are by the pesticides and fertilizers, because pesticides are particularly toxic to insects and they're having a huge effect. I know that from personal experience, because I've been working in this area for about 25 years. I can see the impacts of insecticides on communities of insects at all levels. It's undeniable.

The cost of small drivers in some cases can be important. For example, introduced species: an animal species like possums in New Zealand, which have eaten up all the giant beetles because they are big and easy to spot. This is one of the drivers for this particular group of beetles. Changes in livestock management, in Europe for example, have led to the decline of most of the dung beetles. It's because the dung of the cattle and horses is not available anymore. In the countryside, that dung is collected and used for something else, for manure or just simply buried or discharged in different ways, and now the dung beetles don't have access to it. Therefore, the dung beetles are disappearing, and are the group of insects which is declining most rapidly there.

And there is also to a small extent, some pathogens and parasites that affect some groups, like for example bees that are affected by parasites.

And finally, there is climate change. We don't know much about the effects of climate change, because there are contradictory studies on that. Some of them have shown the increase in the distribution of many species of insects due to global warming. That's very easy to understand, because most insects react positively to warming environments. And the highest diversity of insects is found in the tropics, because it's a warm and humid environment that is favorable for them. At the same time, many species are expanding, particularly in North America and Europe.

In tropical countries, climate change seems to have a negative effect because of a combination of an increase in temperatures and

WE HAVE TO CHANGE SOMETHING. AND THE FIRST THING TO CHANGE IS THE WAY WE DO AGRICULTURE.

declining humidity. Insects need warm temperatures, but also humidity. If humidity declines because it's a drier environment, then insects cannot reproduce and therefore will go extinct. So it seems that in tropical countries, climate change is a negative effect. This is due to two contradictory effects that are benefiting some insects, but at the same time, are negatively affecting others.

PATRICK FARNSWORTH: Your review shows about 41 percent of the insect species have declined or have gone extinct. What is the rate of extinction year to year? Have you been able to measure that rate of loss?

FRANCISCO SÁNCHEZ-BAYO: Yes. We worked it out; it is about 10 percent.

PATRICK FARNSWORTH: 10 percent, okay.

FRANCISCO SÁNCHEZ-BAYO: That has to be taken with caution, because those studies are normally done in small countries like the Netherlands or the Czech Republic, which are easy to study because they are small and there are many experts in those areas, so they can study declines very well. With areas like North America, those studies have been done only in particular states like New Hampshire, Illinois, and places like that. But not throughout the whole country, because it would be an impossible task for anyone.

So, those are local extinctions, but they are still significant because many of the species going extinct or declining are really species that only live in particular areas. If they are not found there, they can be presumed extinct. So in general, the rate of extinction is about 10 percent.

PATRICK FARNSWORTH: That is alarming. So at this rate, how many decades or years do we have? I know you can't predict these things with certainty, and I know it's maybe unfair of me to put this on you. But based on your understanding, how many decades are we looking at before we start to see really dramatic problems, not only loss of insect populations, but our ability to even grow food or anything like that?

FRANCISCO SÁNCHEZ-BAYO: In the paper we hint that within a hundred years we will see almost an extinction of most insects, but as you say some of them will survive. We know that at the same time 41 percent are declining, there are about 5 or 10 percent which are increasing. Those will benefit from the decline of the others.

Not all insects will disappear, but based on the information we have, and studies—like the one in Germany and another one recently in Puerto Rico that measure the decrease in biomass of flying insects, the ones that don't splatter our windscreens anymore because they're gone—we lost about 80 percent now in Europe and Puerto Rico. These are two different regions of the world and it is the same number.

If we extrapolate from that, we can certainly say that within a hundred years there will be less than 1 percent of the biomass of insects left. I know it's only extrapolation. Many things kind of happen: not all insects are going to disappear, some will continue to live in some remote areas. But that's an extrapolation that gives us an idea of the extent of the problem and that we cannot be doing things as we have done them up to now. We have to change something. And the first thing to change is the way we do agriculture, because as I said that's the main driver.

Industrial civilization is killing the planet. Whether we look at it through the lens of climate disruption or through the widespread impacts industrialization is having on ecosystems in countless regions around the globe, we are coming up against the limits of our experiment with an infinite growth economic system. We can't predict the future, but we can identify patterns, and it doesn't take much to recognize where this is all leading.

PART TWO
BLINDSPOTS

DR. WILLIAM REES

Part of the work we need to engage in requires developing the ability to have discussions with people about these subjects, as difficult as that may be. A crucial part of developing the individual capacity to look at this is to recognize how, as a species (and in particular those of us in the global North), we have known about our impact on the biosphere for generations, and just could not be bothered to change course. Why is that the case?

I believe this next segment with Dr. William Rees may shed some light on this subject. Dr. Rees is a human ecologist, ecological economist, and the originator and codeveloper of the ecological footprint analysis. The ecological footprint, as a concept, has become the world's best-known metaphor for the human "load" on the planet, or the resources required of ecosystems to maintain our current mode of living.

Of particular relevance to this discussion is Dr. Rees' work exploring the "neurobiological, cognitive and cultural barriers to sustainability, including human's well-developed capacity for self-delusion." Why are we, as a species, unable to clearly perceive the perilous situation we find ourselves in? Of particular interest for Dr. Rees is how the very nature of large-scale, complex systems (technologies, cultures, societies) enhances the blind spots in human cognition to such a degree that we profoundly and detrimentally impact life systems we rely on for our own survival and well-being.

DR. WILLIAM REES: I'll just give you two examples of what's going on here. And I have to really underscore: we can carry this on for hours, because there are many, many other examples.

Human beings are a product of evolution. We are a species just like all the others, whether we want to acknowledge that or not. And one of the things that evolution has given us is an instinctive response that economists have formalized in the concept of "discounting."

There was tremendous survival value in a kind of short-sighted selfishness. So, imagine a band of hunter-gatherers who are a little bit hungry. Let's say they find a tree with just a few more apples than they can eat right now. I'm using a trivial example, but it's easy to understand. They feed, and there's a few left which they could leave and wander off—they've been personally sated. But if they did that they would not benefit from those remaining fruit. If they move on, maybe somebody else is going to get those fruit. So there's an advantage to taking it with them or over-consuming: having a big feed right now and getting fat and lazy and making damn sure that nobody else acquires those assets.

What we tend to do is discount the future. We tend to prefer the here and now over future and distant places. It's a natural tendency that had huge survival value in the past. It's also what we do when confronted with a decision that would sacrifice pleasures we're enjoying today (such as enjoying my automobile or my big house or whatever it might be) in order to enjoy a possible benefit in the future. And it's only a "possible benefit," because people say the science is uncertain, we really don't know what's going to intervene, and maybe I'll be dead before the future arrives. Why would I do that? Why would I give up something now to gain something maybe in the future? Some unspecified future benefit, such as climate stability?

So that's one thing: we have a natural tendency to prefer the here and now over future events. Economists call this discounting—we tend to underestimate the impact of future costs or, for that matter, overestimate the benefits of our lifestyles today.

Another thing is called the public good problem. Suppose I understand profoundly that my lifestyle is having a three-or-four-fold larger impact on the planet than is fair and equitable, and that is much larger than the Earth could actually sustain—even if everybody else had a much smaller footprint. Suppose I therefore adapted my lifestyle so that I was conforming exactly to what the Earth could handle. I'd have to reduce, say, from six hectares down to one-and-a-half hectares.

So that's it. That's a 75 percent or more drop in the total amount of consumption that I can enjoy in my time on Earth. In doing that, what I've done is made a huge personal sacrifice in wealth. Most of us would see it as a huge personal sacrifice, because we couldn't have our gasoline-powered cars, we couldn't use coal or natural gas to heat our homes and generate our electricity, and so on. We'd be making an enormous sacrifice in order to improve the global environment.

In a sense I'd be creating what economists would call a public good. But in creating that public good, I'd pay the entire cost, but my share of the benefits—assuming nothing else happens—would be unmeasurable. Because my benefit from my cutting back is so small compared to the planet, nobody would even notice. So I make a huge personal sacrifice for an extremely tiny share of any benefit. And by the way, that benefit doesn't get realized anyway, because as soon as I vacate that space somebody else takes it up.

You see what I'm saying? All of this created the space for three or four more people in the developed world, or my neighbor, to get a bigger car and a bigger house and all the rest of it. There's very little psychological incentive for individuals to act now, because they won't gain much anyway—especially if nobody else did anything. And there's always this sense that other people are still pushing and consuming and wanting more and growing and having, and the population is still increasing. So why should I give up anything in order to gain nothing?

So in many respects, we're caught between our instincts to use now, to consume as much as we can in the short time that we're on the planet, and this sense that if we didn't do it, somebody else will anyway.

There's another problem here. This is a difficult one for many people to understand because they think they're living in reality, but we don't actually communicate with reality. Human beings—every culture—socially constructs its perceptions of reality. And then we act out of that reality as if it were true. If you went back say seven hundred years and I said, "hey, the world is flat," nobody's going to dispute it's flat. Everybody can see that. We had a common perception of reality that the world is flat. It's completely wrong, but we all agreed that the world is flat. And you could live your life back then as if the world were flat and it wouldn't make any difference. You'd be dead wrong, but that didn't matter.

The sun rises in the east—we all agree with that. But in fact, the sun doesn't rise in the east. That's a social construct that explains things the way we see them. But the way we see them is a delusion, or an illusion created by our sensory apparatus. The fact is that the Earth is rotating on its axis and in relative terms the sun is in a stationary state. The sun's not rising in the east, the Earth is simply rotating under it. And yet what we perceive is what we take to be real.

That happens in every domain of human understanding. Every discipline in our academic institutions—economics, biology, physics, chemistry—all of these are social constructs. Everything we've learned in those disciplines are social constructs. Every political ideology—capitalism, neoliberal economics, communism—is a social construct, a set of beliefs. Each of them is a separate set of beliefs, values, and assumptions about the nature of reality, about the nature of the economy, about the nature of society, and so on. Religious doctrines—whether it's Judaism, Christianity, or Islam—all of these are stories made up by human beings but believed in profoundly by their acolytes, the people who subscribe to each of those stories.

Human beings make up the realities in which they live. Please understand me: everybody has a disciplinary paradigm, or a political ideology, or a religious doctrine. In fact, we live out of all of those things simultaneously. It's all made up stories that may or may not correspond to the nature of some important aspect of reality. We're a very weird species insofar as we socially construct the realities out of which we act, and those realities may or may not fit the planet in which we live.

To give a very concrete example right now: the world is largely being dominated and run by an economic paradigm called neoliberal economics. It's the economics of capitalism, that the economy can grow forever. In neoliberal economics, human ingenuity is our greatest resource, because through technology we can always substitute for nature or find resources that didn't previously exist, and so there are no constraints on the human enterprise.

That's a model of reality, and it's the one out of which the world economy is operating. Yet that model contains no profoundly useful information about the structure or the biophysical dynamics of the ecosystems with which the economy interacts in the real world. We're operating out of a model of reality that does not map to the biophysical reality within which the economy operates. Then we tend to be surprised the way everything is screwing up.

HUMAN BEINGS MAKE UP THE REALITIES IN WHICH THEY LIVE. PLEASE UNDERSTAND ME: EVERYBODY HAS A DISCIPLINARY PARADIGM, OR A POLITICAL IDEOLOGY, OR A RELIGIOUS DOCTRINE.

I could go on like this for hours, because there's dozens of other reasons why human beings simply don't pay any attention to the nature of reality as our science tells us. It would interfere with the nature of reality that we've created as a means of lifestyle, as an economic paradigm, and so on and so forth.

Human beings are habitual creatures. We've established a habitual pattern that is familiar, that is comfortable. We've socially constructed expectations that everything is going to get better. Then, you map that to the fact that we are natural optimists; anybody born a total pessimist probably wouldn't survive very long. When you tie our natural optimism to socially constructed paradigms of infinite growth and the belief that technology will save us, it takes a huge amount of evidence to make a dent in our actual behavior.

Humans tend to behave much more out of their instincts and their emotional attachments to socially constructed realities. We behave much more according to those things than we do to raw, crude scientific data that says the climate is changing, the fish stocks are dying, or the planet is being eroded under our feet. We just want to shy away from those kinds of things.

I'll give you one more problem, because it really is important. I've already emphasized that we socially construct our realities. What cognitive neuroscience tells us today is that once you've heard a particular idea, or once a particular thought pattern is repeated over and over again, it actually acquires a physical structure in the brain. Habitually thinking a particular way creates neural networks, synaptic circuitry that reflects or embeds in the brain that pattern of thinking. Once you have that ingrained in your mind—and think religious ideology rather than just doctrine or political ideology or academic paradigm—human beings have a natural tendency to seek out other people who think and believe the same way, and to immerse themselves in experiences that reinforce their pre-existing neural pathways.

We learn through the social construction of reality to create reliable social pathways or neural networks in our brains, and then we stick to them. Once we've got these embedded ways of thinking in our heads, we tend to deny, forget, or ignore contrary information. Reason after reason, piled one after the other, reinforces a kind of behavioral conservatism when it comes to facing up to the realities of our ecological dilemma. It has to do with biology, with social paradigms, and with our economic beliefs. And it has to do with this cognitive biology that was once very adaptive. Keep in mind: ten thousand years ago it probably was highly adaptive for an individual to learn the myths, the stories, and the cultural norms of their little tribe. That gave you a sense of belonging and a sense of security. It caused tribal cohesion and social cohesion.

All of these patterns which are currently maladaptive and are driving us to the brink of extinction used to be highly adaptive, because they evolved in a very different set of environmental circumstances. When human populations were small, when technology had no impact, and when humans had almost no influence whatsoever on their environments, it was okay to become entrained in particular set of beliefs, values, assumptions, and myths. It created group security and gave one a sense of identity. Today those same qualities on a crowded planet writhing in agony because of the overuse by human beings have become maladaptive, because it blinds us the seriousness of our own situation.

HOLLY TRUHLAR

To further this point on the limitations human beings have in acknowledging, let alone mitigating, the ecologically destructive impacts of our ways of living and being on this planet (again, this is especially true for the mass societies of the global North), consider the work of Holly Truhlar.

Holly is a lawyer, grief therapist, and community builder, and she caught my attention with her provocatively titled essay "Mainstream Psychology Can Go Fuck Itself," which addressed the ways mainstream psychology largely lacks the framework, language, and tools required to adequately counsel those who are becoming aware of these disintegrating systems.

As Holly states in her essay, mainstream psychology "puts the onus on the individual and takes the focus off of the violent and divisive systems we're in," which in turn disempowers the individual from preparing, in any meaningful sense, for what lies ahead.

PATRICK FARNSWORTH: What have you discovered in your work about human beings that modern psychologists could maybe implement into their understandings of psychology, the human mind, human relationships, etc.?

HOLLY TRUHLAR: I am constantly bringing people back to the ways in which our environment and the systems around us are impacting us. So I would say in a session when it is appropriate, "Well, this is partially due to capitalism, right?" Or, "This is the effects of patriarchy," depending on the client.

One thing I've found is that, more than what we're taught in psychology, self-help, or social work, we are so much more impacted by the collective and the systems around us than we think we are or than what we're told we are. I think that if we're not continually working with that, we're actually doing a disservice.

I'm still really interested in creating more dialog or protocols (but I don't really like the word protocols) around how to respond to people regarding near-term collapse, near-term human extinction. This is why I'm excited about Jem Bendell's work, because what he is doing is actually categorizing: here are possible responses, here's some kind of protocol that I've tried or that I want to put out there. I find that really interesting. I hope that the field continues, and my goal is to continue to do some of that. There's a lot that's unknown.

I actually think most people are more aware of what's happening on a somatic level than what they are able to articulate—even people who, as other people would say, are "asleep" or seem like they don't know. For a lot of people, there's this sort of like buzzing in our bones that we're really aware of, this low level trauma that I think most people are walking around with.

PATRICK FARNSWORTH: That piece does touch on the things we just mentioned: it discusses how mainstream psychology, Western psychology—the profession itself—is not addressing the grief, the despair, the frustration, the anger: all of the things that people are going to feel when they are just paying attention. They're not going to have any the tools, the language, the comprehension, to fit their grief and despair within a Western psychological framework. From your perspective and what you bring up in your piece, why is that the case?

HOLLY TRUHLAR: I think it's multilayered. One thing is that we as a species are ill-equipped for what's happening—I really question whether intellectually we even have the capacity to fully understand, and more and more I don't think so. I think humans have a really hard time even understanding what exponential growth is, let alone the trifecta of what's happening around ecological, economic, and social collapse.

Psychology is actually a pretty young field still. It's more complicated than this, but Freud is sort of the grandfather of psychology, and he's a white man who was working for the upper class. He actually changed a lot of his findings so that he could still get funding, and that has continued throughout the field.

And also—as with most things—it's incredibly affected by neoliberalism. There's this real focus on the individual and individual choices, and there's such a lack of analysis around power, or a systemic analysis or group analysis. On top of this, it wasn't until the last 10 years that we've developed a real vocabulary around

THE AMOUNT OF TRAUMA THAT'S NOW IN THE FIELD OR IN OUR SPECIES IS VERY INTENSE, AND WE REALLY MUST HAVE COMPASSION FOR THAT, WHILE LOOKING AT THE WAYS THAT WE STILL NEED TO BRING OUR ADULT SELVES INTO THE ROOM.

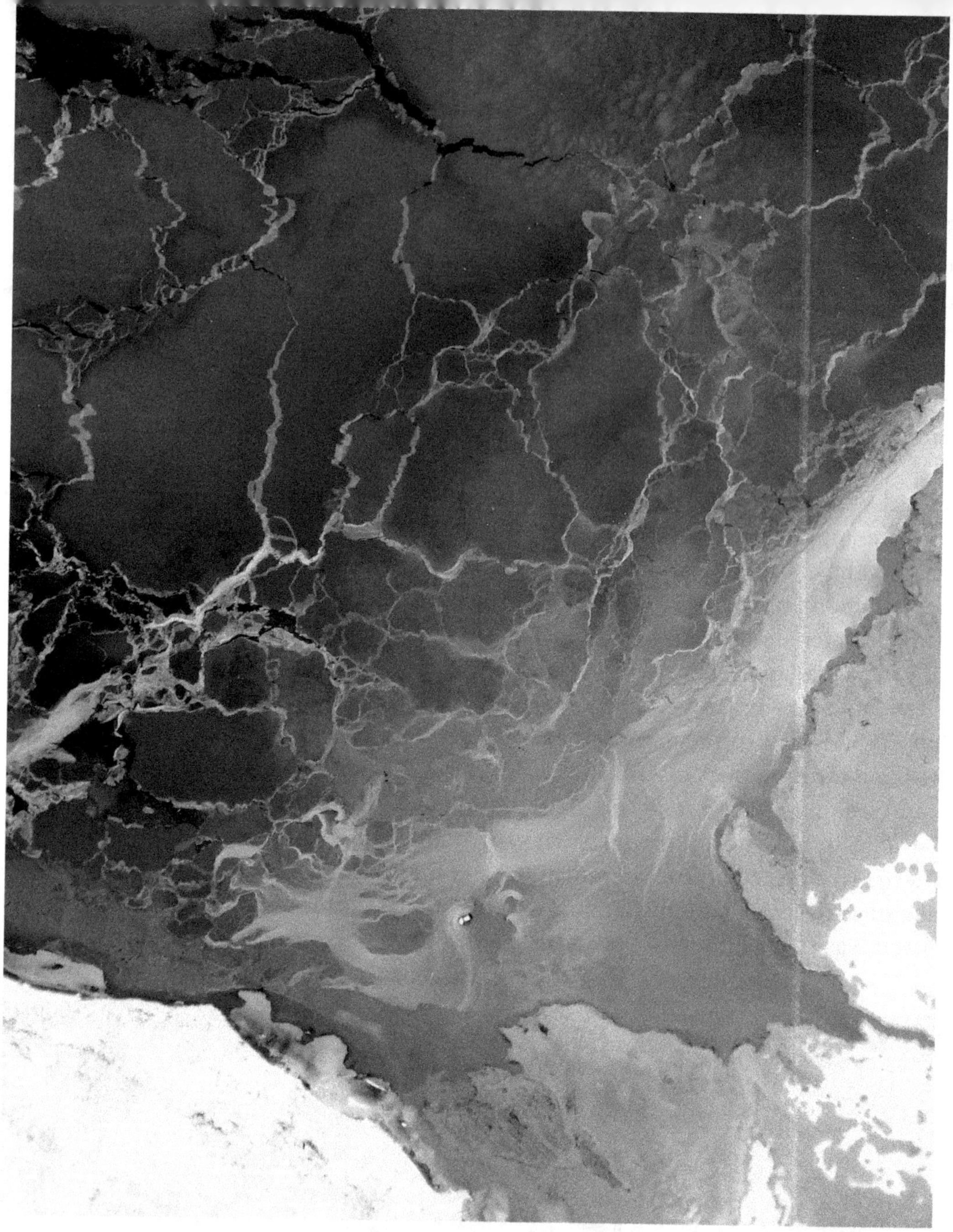

attachment and relationships and what that means. Attachment couples therapy is the new hot thing now. It has roots in attachment parenting and all of that, but the research around it is really new, and it's hyperfocused on couples or the nuclear family. There's still a lack of research and vocabulary around communal secure attachment and what that means.

I think there are also historical roots to what it's lacking. Psychology is colonized. We're really talking about something where the layers we have to go through to find something healthy are so thick it almost seems magical. It seems like we're talking about magic when we talk about the field or the third body that's created when a group gathers or what's being dreamed through a group. That is what I'm interested in, more and more.

IAIN MCGILCHRIST

I would like to further examine the ways that, through a combination of evolutionary processes and cultural conditioning, we have become detached from the living systems of the earth, unable to receive and respond to her signals of distress. It's important to note that not all cultures and societies are as out of touch with living systems of the planet as we are. Our current mode of existence may be relatively new, but the seeds of the dominant culture stem from somewhere, and I think the work of the next individual provides a part of the answer to that question.

Iain McGilchrist is a psychiatrist, lecturer, and the author of *The Master and His Emissary: The Divided Brain and the Making of the Western World,* in which he lays out a detailed picture of the "divided brain."

Iain elaborates on the evolutionary and adaptive perceptual functions of the left and right hemispheres of the brain, and how the balance, or imbalance, of these relatively different perceptual functions has manifested in the trajectory and development of human cultures and societies. The Western worldview (the value systems that have been cultivated in predominantly Western societies), over the course of thousands of years of development, leans more toward a "left brain" form of organized thinking and cultural development.

How has this form of perception informed our relationship with the broader communities of life on this planet? Iain has uncovered a very important piece of this puzzle—how we have gotten ourselves into the precarious moment we find ourselves in today. As Iain has stated, "the mind and brain can be understood only by seeing them in the broadest possible context, that of the whole of our physical and spiritual existence, and of the wider human culture in which they arise—the culture which helps to mould, and in turn is moulded by, our minds and brains."

PATRICK FARNSWORTH: You know, when you talked about the computer and algorithms it reminded me of a conversation I had with a gentleman named David O'Hara, a professor at Augustana University. He talks about philosophy and particularly the ethics of robotics, artificial intelligence, and machine learning. We were discussing was how there's all this emphasis being put on how we can let algorithms take over a lot of the functions of society. Machine learning can take control of cars and how we drive: self-driving cars.

We're starting to see this more and more: in social media and in surveillance, including state surveillance. We see it all over the place. We start to see that logic of computer thinking seep into much of our lives.

What David was very concerned with is that we are putting so much emphasis in this belief that somehow computers are much like the brain. The subtitle of your book is *The Divided Brain and the Making of the Western World.* And that second part, "making of the Western world," is incredibly fascinating to me because we don't have a holistic view of reality, and we're emphasizing so much left hemisphere thinking.

In your view, how did we get to this point? Why are all of these very seemingly clever people—Nobel Prize winning economists—unable to understand that their models for world or national economies are not a real picture of the complexity of the situation? How did the Western world get to this point? How did we evolve to this level of left hemisphere thinking?

IAIN MCGILCHRIST: [laughing].

PATRICK FARNSWORTH: I know that question is ridiculous

IAIN MCGILCHRIST: You've already raised about four different really important points, and each of them would be enough for a day of discussion.

About artificial intelligence, we're just beginning to see how much machines can't do. The big enthusiasm for them is beginning to run into problems, even for doing such simple things as building cars. When you think about making policy or anything like that, never mind understanding a human tragedy, there's no way machines are going to get there. Some say there might be consciousness arising in machines, but we don't know that at all. I'm deeply skeptical that consciousness is something that just arises out of lots of connections.

Just one little example: there are more neuronal connections in your cerebellum than there are in your cerebrum. Cerebrum is the new brain, the bit that we think of mainly as our brain. The cerebellum is the so-called ancient brain at the back of the head, below the level of the brain. It has actually more neuroconnections than the neocortex, yet it doesn't sustain consciousness on its own. You need the rest of your brain for consciousness.

Intelligence is not just something that comes from crunching numbers, from experience, from having a body. Much of what we do in intelligence is actually intuited. Even mathematicians and great scientists describe a lot of their discoveries as being intuitive. That's much more common than arriving at them by following any sequence of steps. You can say, "Well, we could train machines to get more like that." Sure, you can get more and more like that, but you can make a life-like model that can move like a woman and look like a woman and talk like a woman and even smile like a woman—but it still isn't a woman.

Let's clear that one first, in terms of how we got where the hell we are.

I should briefly explain that my book falls into roughly two halves. In the first half, I look at the neurological, the neuro-psychological, and the philosophical implications of the brain being divided. In the second half, I look at the history of the West through the lens what each hemisphere offers to us.

There are three large curves in the history of Western civilizations. I started with ancient Greece, and then the civilization of Rome, and then our modern civilization beginning at the Renaissance. In each case I think I can discern the same shape.

Initially what we find is thinking of the left hemisphere being well combined in such a way as to produce a great efflorescence of learning in every realm, in science, in the arts, in law-making, in philosophy. Each time these civilizations have started with a huge flourishing of everything across the board. And then over time, there has been the tendency for that civilization, as it decayed, to get stuck into the typical left hemisphere's way of thinking: hierarchical, bureaucratic, abstracted. In each case for these three civilizations, it flourishes, then it overreaches itself militarily in terms of empire (be that a commercial empire or an administrative empire or whatever). You can see this in the case of Greece, in the case of Rome, and in the case of the West since the nineteenth century.

The left hemisphere's way of thinking helps you to administrate an empire. It helps you to run a military on a big scale. It helps you grab resources. It helps you make machines. It helps you become wealthy. It is the driver behind grabbing: the left hemisphere controls the right hand with which we grasp things. It also controls the bits of language with which we say, "oh, I've grasped it." In other words, "I've sort of gotten that and put it in my box of things that I now understand."

So the left hemisphere is the one that helps us not to get smart, but to get rich in the short run. But it's not very bright, and in the long run you need balance. The trouble is that the right hemisphere knows that it needs what the left hemisphere can provide: provided that stuff is taken back and used to flesh out the full picture. But unfortunately, the left hemisphere doesn't seem to know what it is it doesn't know. It doesn't know that it needs.

That's where the first part of the title of my book comes in, a fable a little bit like "The Sorcerer's Apprentice." There is a wise spiritual master who looks after a small community in such a way that it flourishes and grows. After a while, he realizes that he can't look after everything himself. More importantly, he realizes that he must not get involved in certain things if he is to preserve his ability to oversee the whole.

INTELLIGENCE IS NOT JUST SOMETHING THAT COMES FROM CRUNCHING NUMBERS, FROM EXPERIENCE, FROM HAVING A BODY. MUCH OF WHAT WE DO IN INTELLIGENCE IS ACTUALLY INTUITED.

So he delegates to his brightest and best, to go about on his behalf and do more administrative procedural things, rather like the way we treat our computers now. But unfortunately, this emissary, bright as he is, is not bright enough to know that he doesn't know very much, that he knows only half as much as the master knows. And he goes around thinking, "I'm the one here who does all the heavy lifting. I'm the one that understands everything. What does he know?" So he pretends to be the master and to do more than he should. As a result, the community falls into ruin, and it's the end of both the master and the emissary.

My message to the West is that we need—very rapidly—to remember things we've forgotten that were once instinctual to us. They involve a bit of humility, but they're very necessary things to remember. We don't actually understand everything, or that controlling things is not always wise. We don't know what it is we're doing when we start interfering. I'm not saying we shouldn't: human beings are such that they do change and control their environment. That's part of being human. But we have to be careful. We have to listen and think about the things that we're neglecting.

Something very crucial was lost on the path to modernity. Whether we trace the roots of the current convergence of crises to the dawn of capitalism or the dawn of civilization itself (or Western civilization more specifically, as Iain points to), we can still, I think, retrieve something of what was lost, even if it won't "save" us.

It's important to reiterate this point: there are things worth doing not because of the outcome, but because they are sacred and true, and ground us to the land we stand on and place us within our right role on this living planet.

RHYD WILDERMUTH

With everything that has been expressed up to this point, I would like to present a segment of my fourth interview with Rhyd Wildermuth. Rhyd has been very influential in how I've approached these subjects since encountering his work through Gods & Radicals, an anticapitalist pagan website and independent publisher, cofounded with fellow writer and friend of the podcast Alley Valkyrie.

Rhyd, through his writings (including his book *All That Is Sacred Is Profaned*, the focal point of this upcoming interview) has pointed to a different way of understanding the crisis of capitalism, by calling upon older pre-Christian, earth-centered spiritual and religious traditions as a means of reclaiming what has been repressed, suppressed, and denied on this path of so-called progress to our present moment.

Acknowledging that there is something like a spiritual dimension to this crisis is to recognize that a truly materialist view does not exclude the realm of the spirit. In fact, the crisis we are in stems, in great part, from the false dichotomy between spirit and matter that has been imposed on us. It cannot be overstated that this is a huge error: material reality is something to be dominated, dissected into its various parts (like a machine) in order to be understood and engaged with.

As Rhyd points to in this segment, capitalism—as a social, political, and economic system—directly informs the ways we relate with the land in which we live, the bodies we inhabit, the labor we perform, the rituals we engage in, and the various ways these relationships and perceptions either highlight or obscure the exploitative qualities of the global capitalism system we are all embedded within.

PATRICK FARNSWORTH: I just visited Ireland and Scotland, and I went to a druid burial site outside of Inverness called Clava Cairns. There was a real—I don't know, maybe I'm just romanticizing it and I'm trying to project all these things onto it—but there was just such a feeling of... home, I guess, but that doesn't seem like the right word.

RHYD WILDERMUTH: No, home is a great word for that. It's a feeling of recognition, a feeling of familiarity, and also a feeling that there is some other part of you we don't use anymore in Western society because it doesn't serve capitalism.

PATRICK FARNSWORTH: Yeah.

RHYD WILDERMUTH: The stuff in my book is trying to scratch at that surface. But this is also Silvia Federici's point in *Caliban and the Witch*, or one of her biggest points. There's so much in that book, everyone should read it. No one does.

PATRICK FARNSWORTH: It's such a good book.

RHYD WILDERMUTH: Yeah. The part where she talks about how things are beaten out of us and disciplined out of us, all of this knowledge that we needed or were familiar with in, say, the thirteenth century.

Where there was Catholicism, there were still relics of pagan practices that continued to exist. Even among some priests, because almost every church around here is built on an old pagan site, and many times you are still able to access the old pagan altar underneath the church's altar. It's completely reasonable to think that some of these priests were still also doing some of the old pagan rites.

If you were around in the thirteenth century, these are the kinds of things you need to know. You need to know that when you walk into this area over here and you have this sense, that means there's something here so you should leave something as a thank you. Or you should not trash it the place, because you'll anger the spirits that are there. But none of that knowledge is useful in capitalism. Capitalism doesn't need or want us to do that stuff.

One of the things that really turned me to the intersection between all of these indigenous pagan or pagan animist beliefs and resistance to capitalism was a New York Times article about the Neak Ta. The Neak Ta are ancestral land spirits in Cambodia, and people still worship them and still bring offerings to them.

Factory owners were building new textile mills all over Cambodia. Everybody was moving their textile manufacturing there because it was cheap labor, the exact same thing that happened in England in the 1700s: they just keep moving these things around. In order to build these factories, they had to displace the ancestral spirit and whatever the spirit's house was. Oftentimes the spirits were attached to trees. It's a bit like Shinto, where the trees house a spirit.

So they would displace the spirit, and then weeks later, the workers, almost all of them female, would have possession experiences where they would start screaming. The workers would say that what was happening was that the Neak Ta, those spirits, were possessing them because they were angry. A bunch of these workers started channeling the spirit, telling the owner of the factory that the factory was going to burn down unless the factory owner met its demands.

For the capitalist, that ancestral belief is absolutely, utterly inconvenient. They don't want their workers to be able to have a sense of, "oh, there's something here, I should interact with it." But I think all of those ideas are still latent in us. When you went to Scotland, when you went to those cairns, you had this sense that was like home, a familiarity, a kind of recognition. But it was also queer in the old sense, not in the gay sense, but in the sense that there's something strange here, something weird here.

That's a sense humans have always had and it's something that Western civilized people don't know how to deal with anymore. Some people turn it all "new age" or whatever, like, "oh, maybe there were aliens here." They make up all kinds of things. But the sense is still there, and the capitalists have never really fully been able to eradicate it from us.

PATRICK FARNSWORTH: It seems, though, that just like anything, capitalism wishes to commodify that genuine feeling and desire to connect to your ancestors or to connect to a spiritual or magical thing. You and Alley Valkyrie discuss this in an episode of your podcast, *Empires Crumble*, and you've written about this extensively as well: the commodification of the "instawitch," the hashtag Instagram witch, or the Sephora witch kits. You buy these little smudge kits, and of course, it's never going to be the real thing; it's a constant redirection of people's curiosity and energy towards buying commodities, or going to Sephora to buy witch kits instead of actually delving into something far more mysterious that may pull you out of a capitalist logic and a consumer logic.

RHYD WILDERMUTH: Right. I wrote an essay on this a couple of years ago, called Paganism™. That essay starts out with an experience I had with two friends of mine, Peter Grey and Alkistis Dimech, who are the publishers at Scarlet Imprint. We were in this ancient stone circle on top of a mountain in Wales, or actually right on the border of Wales and Shropshire, England.

It was a profound experience to be there, like, "holy shit, this thing is ancient and also a little angry." It was an angry site. You go there and you feel not really a malevolence, but more like rage at something. I've come to conclude that this usually means people were murdered there. I suspect that there was some battle where druids or something were slaughtered. Oftentimes I'm correct. Sometimes I have no way of backing that up, so I've no idea.

So we're there, and it's this old worn down stone circle. Peter goes up to one of the stones, and he picks up a bunch of stuff that's at the bottom of the stone. And I'm looking at what he's picking up: a bunch of coins and crystals and a pink plastic fairy. My initial thought was, "oh, that's kind of cute. They don't really know what they're doing, but people are trying to put offerings here."

Then Peter throws the stuff angrily and called it something like "fucking neo-pagan trash." I was standing there, trying to reconcile this feeling that it's cool that people are seeking things—which is this sort of coddling New Age-y kind of thing—with the rage of this guy who's my hero. Why would he do that?

THAT WAS ALWAYS THE BRILLIANCE OF WHAT MARX WAS DOING, POINTING OUT THAT WE HAVE TO LOOK AT THE MATERIAL CIRCUMSTANCES THAT CREATED A THING, THE CIRCUMSTANCE THAT CREATED THE SITUATIONS WE ARE IN NOW. IT DOESN'T LET YOU PUT A LEVEL OF IDEOLOGY OR A FANTASY OVER IT.

I actually tried to explain to him, too. I said, "it's kind of cute that at least people are trying to do what pagans used to do before—they just don't know what they're doing." But then out of nowhere, I get this image, totally unbidden. I don't know where it came from. It was an image of Chinese women in a factory, stamping pink plastic into the shape of fairies.

And it was at that point I understood. There is the raw ancient thing versus a neopagan capitalist petrochemical that was dyed pink with who knows what by low income Chinese workers, who worked 12 to 16 hours a day to make these shitty little plastic fairies so middle-class white British people can buy them in shitty little occult or pagan stores and leave them on ancient standing stones where there used to be sacrifices and druid circles.

PATRICK FARNSWORTH: I see what you're trying to present there. If I were a land spirit, if someone brought me something like that, I would be like: "I see what you're trying to do, but it's not what you think it is." Because as you mention in your book, the animist perspective on matter, or reality or the world that we're all in, includes within it that plastic fairy made in a factory in China. And it has within it the very emotional and spiritual quality of how it was made. If it's being made by exploited low wage workers in a factory, and it's being shipped all the way around the world to be sold to middle class Europeans or Americans so they can fulfill their fantasy of being a little witch or druid for the day, that trinket has that in it.

It's not going to go away just because you have a good intention of putting it on that stone in that sacred site: everything is interconnected. So, in fact, the very thing that I'm using to talk to you right now is made with various minerals extracted from some mines somewhere, probably...

RHYD WILDERMUTH: ...by kids in Africa.

PATRICK FARNSWORTH: Exactly. And these hidden layers are invisible—invisible layers of exploitation that go into giving us this fantasy world that we're all in because it serves a capitalist market.

RHYD WILDERMUTH: You know, we can actually segue to the book here. In the third chapter I introduce what is meant by Marxist historical materialism, the materialism itself within Marxism. For a long time communists saw materialism as being an atheist position, because superstition gets in the way of talking about material circumstances. But then there's animism, which is ultimately a materialist philosophy because animists don't see a distinction between the material and the spirit.

Animists look at a thing and don't divide it between material and spirit. For instance, a tree is not just a thing. It's constantly expanded by its existence and our way of seeing it, but you never forget the material circumstance of the tree as a living body. That same animist materialism helps you understand that pink plastic fairy or the cell phone. Those things hold the sorrow, the suffering, and the exploitation of the workers who created it. The labor that went into that doesn't go away.

That was always the brilliance of what Marx was doing, pointing out that we have to look at the material circumstances that created a thing, the circumstance that created the situations we are in now. It doesn't let you put a level of ideology or a fantasy over it. Going back to those pink plastic fairies: the fantasy says it's an offering to the fairies but materialism says, "it is petroleum plastic that has been dyed pink and shaped by poor Chinese workers who were being exploited. That is what this thing is."

You may be using it as an offering to the fairies or whatever, but in actuality your offering to the fairies is created by exploitation. And so you can't deny that whole thing. And you can extend that to absolutely everything else. For example, the abortion argument in the United States: the official reason why people want to stop abortion is "because we believe that life begins at conception." That's an ideological and a religious overlay. When you look at the material circumstances, and the material need the rich and the government have to prevent women from shaping their own reproductive activity—whether or not they have a child—you see cheap labor. The more poor people you have, the lower wages go. The lower wages go, the higher the profit is for a capitalist.

This materialism, both animist and Marxist, is just looking at the longer chain of circumstances, rather than accepting the ideological and fantastical overlay that the people oppressing us put on the world.

There are hidden relationships that bind this world together, that bind us to the earth and to each other. The animist perspective acknowledges these relationships, respects them, names them, and encourages practices that honor them. The food we eat, the clothes we wear, and the land we live on—our bodies extend beyond our physical forms. The processes, both on the micro and macro scale, that produce the material representations of our relationship to life on this living Earth need to be seen for what they are. The nature of our relationship to these processes inextricably ties us to our collective fate, for good or for ill.

PART THREE
HUNGRY GHOSTS

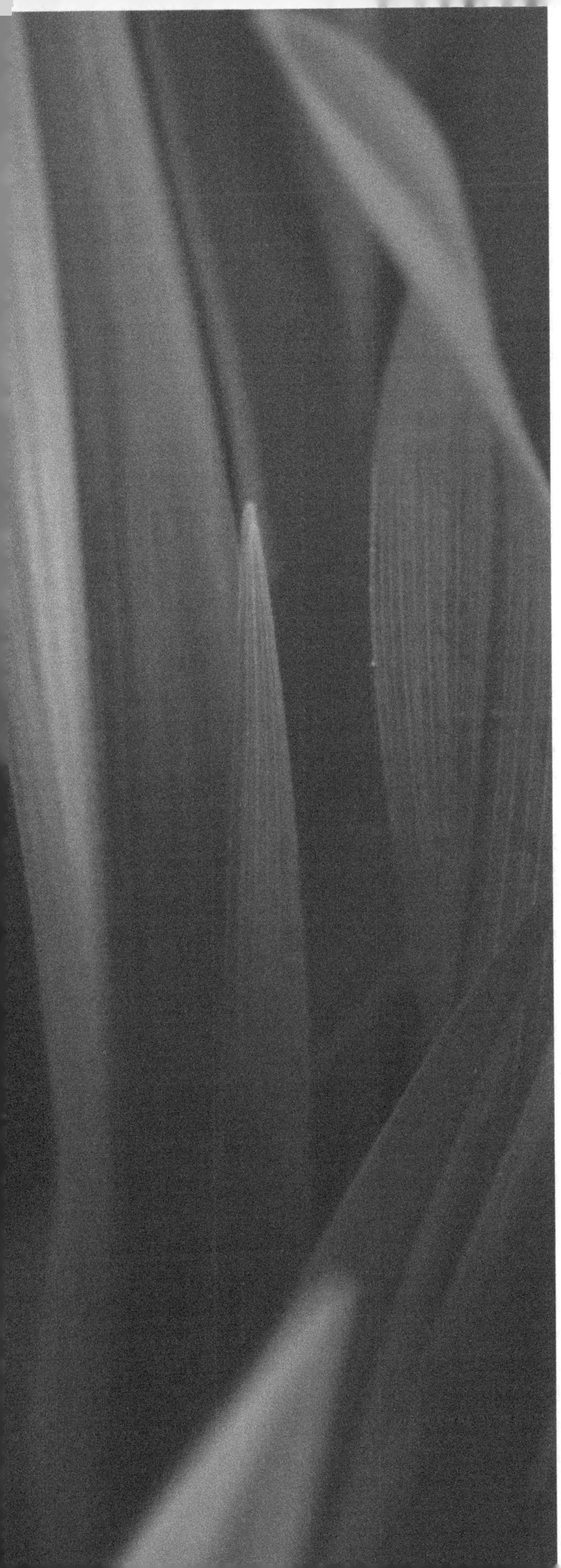

IT MAY BE CONVENIENT and logical, at this point in this book, to come to the conclusion that we were always going to reach this point of no return, that we are doomed to our own widespread self-destruction. Based on the science of where we are currently at and projected to be in the foreseeable future, this is a completely understandable position to have.

We need to dig a bit deeper, though. Developing an expansive view of the variety of human cultures and social arrangements that have existed reveals that the reality of our current predicament is far more subtle, contradictory, and complex than we often assume. There have been many potential futures; the future we have before us currently portends our own potential extinction and the extinction of so much of the nonhuman life on this planet. In spite of this, the future we are carving out for ourselves was not inevitable, not in the slightest.

SILVIA FEDERICI

I've mentioned Silvia Federici's work and perspective, in particular her book *Caliban and the Witch: Women, the Body and Primitive Accumulation*, numerous times in my podcast throughout the various interviews I conducted (for instance, in the previous segment with Rhyd Wildermuth).

Her work has been deeply influential in furthering my understanding of how we have arrived at this point of converging crises in all realms. Silvia's astounding work has provided a crucial analysis of the various historical processes and material conditions that have led us to our ecological crises, our political crises, and our crises of the self—our alienation from our bodies and from the land.

We often lack the necessary historical context to understand that these crises didn't come about because human beings are naturally brutish and warlike, or that it was inevitable that we were going to despoil the entire planet in the name of economic growth.

For a long while I had this misanthropic perspective, that human beings themselves were the problem, a virus or disease on the body of the earth. But coming across Silvia's expansive work and research into the witch hunts, colonization, the enclosure of the commons, and learning how these processes led to the violent birth of the global capitalist social order, I realized there were specific historical paths we took to get where we are now.

Silvia explains what function the witch hunts served in Europe during the medieval era, far beyond the superficial understandings we often have of it. This phenomenon in which primarily women—mostly poor women—were targeted, tortured, and publicly executed paved the way for the rise of the world-eating socioeconomic order. During the witch hunts, tens of thousands, potentially hundreds of thousands, of women were killed, and this event is rarely given the attention it really deserves.

Silvia, being an autonomous Marxist feminist, recognizes the value of digging into this history, examining how it ties into other trends that were happening simultaneously. We need to understand that it started with women, and that capitalism emerged from the subjugation of women and their central role in community and radical resistance to the feudal order that preceded the birth of capitalism.

PATRICK FARNSWORTH: What you're explaining is that the wealthy—those that had the power in this feudal society—were seeing their power could very easily be taken away from them by the organization of large groups of peasants. Those peasants were willing to wage war against them in order to reclaim their freedoms and their liberty, and they understood very well what that meant.

Something that you go into in this book is how this affected women and women's role in those societies. This transition was not an evolution, because like you state, it wasn't like a natural evolution that capitalism evolved out of feudalism. It was a very calculated and thought out; the intention was to move in this direction, to put the power back in the hands of the lords, the people in charge of the land, in charge of the economic system. What happened to women in this process?

SILVIA FEDERICI: One connection that I've made is that capitalism is a system of power, it is a system that affects every form of life. So one of the things that I've been trying to look into, particularly in mind of what happens to women in capitalist society, is to understand: well, what's the difference?

This previous system, the organization of material lives in feudalism, what was the position of women there? Certainly the position of women was not one of equality with men, but I found that the level of inequality in the servant, peasant class between women and men was much less pronounced than it is in capitalism.

A person that serves in the peasantry was referred to within the feudal system as bondsmen and bondswomen. The ultimate authority on the family, on marriage, and basically family life was in the hands of the landlords, the feudal lords. So you have less the kind of indirect rule over women than you have in capitalism. In capitalism there's a very clear delegation of power that serves capitalism: you have the state and capital delegating power to men, power to the wage worker over the unwaged housewife.

In feudalism, you have a situation where both women and men have access to land, have access to means of reproduction, but where power relations, for example marriage, are very much in the hands of the lords. A landlord could oblige a woman who wished to remain a widow to marry again. So the difference in power between men and women was much more limited. And the fact that women did have access to land meant the women could participate in their own production.

In feudal society, the separation between production and reproduction is almost nonexistent. This separation comes into being in capitalism, where you have production for the market, production for the consumption of the family, for the consumption of the worker, and for the reproduction of the worker. It's only in capitalism that you begin to have this separation of two spheres of life. And as we know, reproductive work almost disappears as work.

In the feudal system you don't have this. In agricultural work, the family produced and cultivated on the plots of land that were allocated to them. When the community cultivated the land, there was production and reproduction at the same time. You do have a certain amount of produce that is sold on local markets, of course. But people are selling the excess, wherever they have excess, and often it is bartered. Nobody in that kind of society depends on money for their survival; in fact, money is much more limited, very limited: even the wages paid are limited in nature.

You begin to have a separation between production and reproduction at the moment when workers do not any longer have any access to the means to reproduce themselves. When they've lost their land, they depend on wages: whether they're working for wages or not, everybody depends on money for the ability to reproduce themselves. That's when you begin to have these two different spheres, production and reproduction. In the feudal period, you don't have the separation between man being the wage earner and producer, the one who gets the wage, and the women depending on them for their sustenance because they don't have access to it. You don't have that. Everybody's working with the land.

I've also shown that there is a strong presence of women in the social movements, and particularly the so-called heretic movements, which appear to us as religious movements because they used a religious language. Actually, these were social movements. They were movements for social justice, because they were protesting against inequality. They were protesting the exploitation by the church. The church basically tried to use its power to make money selling baptisms, selling indulgences, and basically using all the religious liturgies, services, as a way of making money.

So the heretic movements were in a way an "another world is possible" movement. They would say: "Christ was poor." Christ did not have property, why should the church have property? Women were very central to this movement. And in the heretic movement, women were also allowed to participate, to administer the sacraments.

The third point is that the idea of the witch begins to develop out of the last phase of the struggle of the church against heretic movements. The idea of the witch begins to be forged in a religious environment. The witch hunts actually take place much later, when they become a mass phenomenon: mass persecution. It's in a much later period, and it's organized not by the religious authorities or the clergy, but mostly by magistrates, by the local political authorities.

I'm trying to show that there is very much an agenda here. It is very important to understand why in this period, particularly in the fifteenth century, around 1400, 1450, you begin to hear about witches. You hear about witches and people begin to write treatises on witchcraft. At first, it's a very small phenomenon, then much later it becomes something much bigger, becomes a large persecution. This is the kind of context that I have tried to reconstruct, to understand, to have a much fuller idea of. What were the forces? What was the terrain in which capitalism took off?

PATRICK FARNSWORTH: So the witch hunts are a very dramatic phase, and it loses this dimension when it's talked about in history textbooks—it loses its historical context. Also, the transition to capitalism is not discussed as the reason why the witch hunts occurred in the first place.

The people that were trying to capture these women accused of being witches, labeling a woman as a witch and then executing her publicly: what was the purpose of that? And what were the implications of the witch hunts on women and on the development of capitalism?

SILVIA FEDERICI: It's like when you look at a landscape, and the farther up you go, the more you'll get a comprehensive view. The moment you zero in or zoom in all you see is a street. But when you see the street in the context of everything else, you'll have a much better view. I'm using this for the witch hunts, because it's very easy to get lost in them, lost in particular accusations or in a particular sense of persecution. But then when you look at it as a whole, then you get it.

"As a whole" means both in time—over the period of two-and-a-half centuries in which these were particularly massive—and looking at all the countries involved. When you put it into history and what else was taking place at the same time (the conquest of America, the beginning of the slave trade, the expropriation of the peasantry, the beginning of a capitalist system), then it begins to speak to you in a different way.

I've been trying to stress that in every place across Europe, it begins with governments passing laws—a new crime: witchcraft—and new regulations, new acts of government. This is very important, because it is from above that people are told "watch out." It's not something that is coming from below.

There's more that happened than what has gone into *Caliban and the Witch*: for example, one thing that I could have put into the book is that a poor person in the sixteenth and seventeenth century, going to a magistrate to accuse a woman, would not be necessarily listened to. Who is being listened to, whose accusations are heard and whose are not heard already tells us the class nature of the witch hunts. Also striking is the difference in class between the so-called witches and the accusers. There's a class difference: those who are accusing them are usually the people of power in the community, people who have property. It's a persecution of a very clear class character.

OUT OF THE WITCH HUNTS COMES A WHOLE NEW DISCIPLINARY REGIME THAT BECOMES THE NORM FOR WOMEN IN CAPITALISM.

And there are different groups of women: there are differing groups of women who are vulnerable to the accusation. It's not one group only, but there are common denominators. They are generally the poor. They come from the lower classes, and there are certain types of accusations that are very common and often piled up: a witch is accused of all of them at the same time. So I tried in *Caliban and the Witch* to give a typology, and that typology of the witch is also a typology of the changes to capitalism, particularly in the sphere of reproduction and in the sphere of relations between women and men.

The witch is often the beggar, the old woman not accepting that the people refuse to support them. There's an attack on communal mutual aid, an attack on the idea that people have a right to support themselves. And there's a fear by the better-off about the poor, the poor that have been impoverished. Because why should there be women who don't have anything, who have to go around the neighborhoods asking for some wine, some oil, some butter, or who are living alone in poverty?

There's also women who are exercising healing practices. Basically they're midwives, or they're curing people with herbs, sometimes with charms. Certainly many women had their own gardens, and they knew the properties of plants, they knew the properties of roots and flowers, and they were the doctor. So these are women who have a certain power. Often they predicted the future. So the figure of the healer represents something in the community. And so you have an attack on popular powers.

You have an attack on mutual aid, you have an attack on popular powers. Then there's also the woman who is supposed to be promiscuous, right? There is a whole campaign on the question of sexuality and procreation. The witch hunts are also a way for the state and capital to begin to exert a new type of control over the body of women. They controlled procreation, to make sure that procreation is productive, to make women use their sexuality in a productive way, to give birth. Also, sexuality has to be controlled: sexuality is seen as something that subverts social order. It can subvert the people's relation to work. It can subvert the class differences. So the body of women begins to be portrayed as something that is dangerous, a place of dangers that have to be neutralized.

Out of the witch hunts comes a whole new disciplinary regime that becomes the norm for women in capitalism. There is an important difference between the image of the woman before the witch hunt in the popular literature and the image of the woman as it appears in the literature of the seventeenth and eighteenth century after the witch hunt. The witch hunt is a defeat on the power of women, an attack on the social power of women. And it's also a preparation for women to take on particular tasks. She becomes the unpaid worker, she becomes the one confined to the work of reproduction, and she is portrayed as not having much reason, being weaker without reasoning power and thus needing to be controlled by men. And so the witch hunt is really the condition or the step towards the creation of the new sexual division of labor.

This interview with Silvia, and her book Caliban and the Witch, really illuminated for me the processes that led to the emergence of capitalism, and how the witch hunts and the enclosures of the commons were crucial but often overlooked events that contributed greatly to the rise of this new social order imposed from the top down. This sustained violence against women in particular contributed to our severance from the land and from the body. The social divisions that emerged out of this informed the rise of the global capitalist system, and in turn, the deeper disconnection we have from our ancestral knowledge (especially as settlers) and the living earth as a whole.

DR. GERALD HORNE

For more on the historical roots of our modern crises, consider next my interview with historian Dr. Gerald Horne, author of the book *The Apocalypse of Settler Colonialism: The Roots of Slavery, White Supremacy, and Capitalism in 17th Century North America and the Caribbean*. Dr. Horne's elaborates on the formation of the trans-Atlantic slave trade, the formation of the American colonies, and how the United States itself is built on the ideas that emerged the same time that the witch hunts were occurring in Europe.

As European nations began to colonize and carve up much of the world, what they did in Africa, the Americas, and elsewhere was part of the same process as had been done to women and to the peasant class in Europe during the witch hunts and the Enclosure Acts.

The Enclosure Acts were enacted across Europe as governments began closing off the commons from the poorest segments of the population and converting that land into capital and private property, setting the stage for the "apocalypse of settler colonialism" in the so-called New World. It's not a coincidence that in the wake of the witch hunts, the genocide of the Indigenous populations in the so-called Americas and the mass enslavement of Africans occurred to meet the burgeoning economic demands of the early capitalist economy. As Dr. Horne explains, the slave trade is a crucial component in this transitional process, speaking to the foundations of the economic prosperity of the Americas as a whole, and in particular the United States of America.

Many of the problems we have now come from a lack of understanding of our own history. The ghosts that linger on this land—the mass genocide of Indigenous populations, the ever-present impacts the slave trade and colonization have had on the African populations—were founding victims to the formation of the social construct of race. The idea of being white, black, or any other race is rooted in the construction of a slave economy, which in turn gave rise to the capitalist economy we exist within today.

DR. GERALD HORNE: I'm doing this project on the 1500's right now. One of the things that strikes me particularly with regard to England was that England was under tremendous pressure from the Spanish, who had first mover's advantage because they sponsored Columbus's journey. You may recall that it was in 1588 that only a bout of bad weather kept the Spanish Armada from conquering England. And as a result, England tried to curry favor with black people, because as you know England doesn't move into the Americas until the 1600s and I'm talking about 1500s. So they're currying favor with black people in order to counter Spain.

For example: Panama, which had a heavy black population, Sir Francis Drake, Her Majesty's pirate, curries favor of black people when he arrives in Panama in the late 1500s. Also, in Morocco: England has this trade relationship with Morocco where it sells Morocco weapons and then gets back saltpeter to make ammunition. And even in the early 1600s there's a kind of queasiness in terms of some of the settlers enslaving Africans. Not necessarily queasiness about enslaving Indigenous populations, but certainly about enslaving Africans, and that begins to change around 1655.

So what I'm trying to grapple with is that age-old question: does slavery help to give rise to racism and white supremacy, or is there a pre-existing white supremacy that gives rise to slavery? I have to say I'm biased towards the former idea; that is that slavery helps to give rise to racism. I'm trying to maintain an open mind, because I recognize that early on you have the Portuguese going into West Africa, enslaving Africans and bringing them back to Europe as early as the fourteenth century, the 1300s. You have certain Iberian nationals going into Africa and trying to set up plantations on African soil, like São Tomé e Princepe, this island off the coast of West Africa.

But then again, you look at the work of the great classicist Frank Snowden, *Before Color Prejudice: The Ancient View of Blacks*. He tried to bring forward evidence to suggest that there is no necessary animus towards darker skinned people. And Miranda Kaufmann in her book *Black Tudors*, which deals with the 1500s, was very interesting. She talked about the black population in England in the 1500s and cites this interesting case of a black man who beats up this man who you would define as white, who of course is English. And there is no necessary recriminations or repercussions. And she brings this forward as evidence to suggest that

AS EUROPEAN NATIONS BEGAN TO COLONIZE AND CARVE UP MUCH OF THE WORLD, WHAT THEY DID IN AFRICA, THE AMERICAS, AND ELSEWHERE WAS PART OF THE SAME PROCESS AS HAD BEEN DONE TO WOMEN AND TO THE PEASANT CLASS IN EUROPE DURING THE WITCH HUNTS AND THE ENCLOSURE ACTS.

there is no pre-existing animus towards darker skinned people.

I'm grappling with all these ideas and, as I said, I'm predisposed to think that slavery gives rise to racism and white supremacy but I'm trying to maintain an open mind.

PATRICK FARNSWORTH: So how did slavery ultimately inform the socioeconomic system that we have today? Do you think that we would have such a developed a form of capitalism without slavery? Or could we have reached this moment without the enslavement of countless human beings?

DR. GERALD HORNE: I would tend to lean towards the latter point of view, that the great wealth that seems to exist the United States is heavily grounded upon slavery and the African slave trade. Keep in mind that with the African slave trade you begin to see the enhanced construction of ships, not only to transport the Africans across the Atlantic but also warships to batter West Africa and East Africa into submission in order to gain these enslaved. And also to beat off rivals: the Dutch, the Danes, the French, the Portuguese, and the Spanish.

Then, there is the question of financing these voyages through banks, which gives rise to a banking system. And of course you need workers to construct the ships, and they are getting wages in western Europe and are able to buy other items: that leads to spin-off industries there. And then there are insurance companies which even today control huge amounts of capital because there was always the danger of shipboard insurrections by the Africans—that had to be insured against. And so you begin to see this take off of the system we now know as capitalism in the seventeenth century, precisely because of slavery and the slave trade.

I should say that I try to take this argument forward, but it would be a bridge too far to say that I'm the only person who's thought of this idea. I mean: Walter Rodney and his book *How Europe Underdeveloped Africa*, Eric Williams and *Capitalism and Slavery*, the University of Rochester historian Joseph Inikori with *Africans and the Industrial Revolution in England*. Many other scholars have dealt with this question as well, and it's well recognized that in 1860 on the verge of U.S. civil war the most valuable property in these United States of America was enslaved Africans. This leads to the conflict over slavery that causes this great civil war, leading to the deaths of seven hundred to eight hundred thousand people.

And then the fateful decision is made post-1865 that not only will the enslaved population not be granted reparations, but also the slave owners won't be compensated because they were compensated when Britain abolished slavery in the 1830's. And when the slave owners in Dixie were not compensated, that leaves many of them poor with their property—which previously they had come to see as not part of the human family—walking around claiming equality. This helps to engender wrath and fury, leading to the rise of the Ku Klux Klan and racial violence, racist terror, and perhaps bruised feelings. This country has really not come to grips with that particular decision, not to mention everything else that we've talked about.

So I guess the long and short of it is this: there needs to be some sort of reckoning in the twenty-first century with regard to these issues that historians like myself are raising.

The concept of race and the identities that have formed around this socially constructed dynamic have particular relevance when we try to understand the multilevel crises we are currently experiencing. To explore one particular branch of this, I want to present a segment of my discussion with Tad Hargrave.

In my interview with Tad, we discuss the difficult subject of race, in particular the modern and the historically recent formation of the concept of "whiteness." To be labeled and exist as a "white" person in our current era is to live with certain privileges within the dominant culture of North America, some of which are subtle, and many of which are not. Along with the privileges that come with existing as a white person in modern North America, to be white is also to lack a connection to the land and place that white people reside in. Also, to be white is to bear unacknowledged wounds: intergenerational traumas that inevitably stem from being the descendants of people who fled the brutalities of Europe during centuries of land enclosures, colonial expansion, and rapid industrialization. In the process of assimilating to the dominant culture of North America, North Americans of European descent shed their ethnic and cultural traditions and ties to their homelands, adopting the impoverished racial identity and social category of whiteness.

Of course, I'm speaking in broad generalities when it comes to this subject, but understanding the historical trends and events that have brought us here can help us have honest conversations about the subject of race and to reckon with what we may find in this exploration.

PATRICK FARNSWORTH: I grew up Mormon, and I don't know if you're familiar with the LDS church, but it's a product of American culture. It's fascinating when you step outside of it, and look at it from the outside in. Mormons are incredibly fascinated with genealogy. They want to know where their family came from. They want to know their ancestry. They do a really good job—they're some of the best at it. But to me, it almost turns into this fetishism: how much Irish do you have in you, how much Scottish, how much Danish, how much this or that, all without actually recognizing why they come here in the first place. Why didn't people decide to stay in their homelands? What series of events pushed them out of the place that they called home to what would be called now the United States of America or Canada?

The title of your blog, *Healing from Whiteness*, is really interesting to me. I think it's important to discuss what it means to step away from this identity of being white, because it was something that we're born into and we don't really fully understand. It's a form of disconnection. So I want to get your thoughts on what it means to heal from whiteness.

TAD HARGRAVE: I don't even know if I'm attached to the blog title. [laughing] Give it a few years if it's still a good idea. I suppose what I'm trying to point to is that whiteness was imposed, or taken on out of desperation.

It's an understandable thing if you're a starving Irish mother and you've got to feed your kids. It's not like there was a conversation of, "okay, now become white and we'll give you these things." But over time you get the idea that my kids can have a better chance if they don't speak Gaelic, if they're not easily identified as Irish, because there's such a prejudice against the Irish. So you don't speak the language to your kids. It's little things like that: a lot of the ancestry is missed, and so much gets lost.

I would say whiteness is economic political privilege, but it is also spiritual and cultural poverty. And it's in vogue to talk about white privilege, and I'm not disputing it, but there is a poverty that comes with it. It's very easy to miss that. There's this hunger and this sense of, "you know, the place that I come from is just the home of colonizers." And that's all we've ever been and are. That's pretty terrible. And yet I still feel this aching spiritual cultural hunger in me, and so I'll go and get my fix from brown people from their ceremonies because that's where the real stuff is. Those are "real" human beings and white people never have been, and then whiteness gets projected back in time.

Whiteness comes from a certain time, but then people will say "oh white people have always been like this." But hold on: they weren't white for certain dates and times. This comes from a certain time and place, mostly North America. I don't know if you've heard of Steven Pinker?

PATRICK FARNSWORTH: Oh yeah. Definitely.

TAD HARGRAVE: Yeah. And Jordan Peterson. Jordan Peterson has been getting a lot of hype lately—a number of my friends were very into him and Steven Pinker. Understandably, because he's spreading a lot this hope mongering—look how good things are, it's so much better than people think it is, everyone's being too pessimistic but there is less violence, less war, all of that. Fundamentally what he, what Jordan Peterson, and what a lot of others are making the case for is Empire. They're making the case for Western civilized life. Which is, "Isn't the Enlightenment amazing? Look what Europe has done for the world!"

This is the exact same as the white-hot center of white nationalism: look what Europeans have done for the world. Of course when they say Europeans, it's a very specific thing. What they mean is Empire. You've never heard a white nationalist say, "You know Europeans done a lot of good for the world, like my ancestors in the Appalachian Mountains we grew apricots and had 40 varieties dried on our rooftops—it was amazing!" Or, "You know, my ancestors in Scotland invented a foot plow, it's an incredible idea!" Or, "You should hear our folk songs and stories, some people would know hundreds of stories!" It's never any of that. It's always roads, or the industrial revolution.

Clearly, the world is in a bad place, and when we look for the solution within Empire we get more Empire. But our whole identity is rooted in Empire—which is what whiteness is. That's one thing I said right in the beginning: whiteness is one of the most recent branches on this tree of Empire. With white nationalists and people like Steven Pinker, we got to be proud we're white. But that's the spell in full effect: we're defending that which afflicts us.

Somebody I studied with has a really beautiful formulation: "first generation trauma, second generation God." Basically, the thing that traumatizes a generation of people will likely be deified by the next generation of those people. So the English come into Scotland with colonization, and the first generation will resist. But the second, or maybe the third, will want to be English. It's not uncommon, and it's not always so overt. When I was going to the Gaelic College on the Isle of Skye, Sabhal Mòr Ostaig, I remember somebody saying "we're just as good as any English university." As if that's the basis, as if we need to be comparing ourselves to that! But it happens like that.

So the thing that was so destructive to our old timers? We've now held it up as the cure, as the medicine that we need, when it was really the sickness that came in. It's like people getting measles and then saying these red spots are a sign of my health and virility and refusing any treatment. This is very hard for a lot of people to hear, that whiteness could be a sickness of sorts. And it's not the European descent that's the sickness, because there's so much beauty in traditional folk and Indigenous Europe. But whiteness is a different thing. Whiteness is like the shroud on the mummy. It covers, so you never see the body.

SHANE BURLEY

Continuing on the subject of race and especially whiteness, here is a segment of my third interview with Shane Burley, journalist and author of *Fascism Today: What It Is and How to End It.* This interview was recorded just days after the mass shooting in Christchurch, New Zealand in March 2019, in which an avowed white supremacist entered two mosques and killed 51 people.

Prior to the shooting, he released a manifesto laying out his motives for the attack. The text is steeped in racial identity politics, with an ecofascist bent thrown in for good measure. Many of the convictions of these "lone wolf" terrorists, the Christchurch shooter included, are steeped in pseudoscientific notions of the supposedly inherent differences between the races, even though these ideas have been largely disputed by serious inquiry and research for decades. This is especially relevant when we take into account what Dr. Gerald Horne has elaborated in his work, that our notions of race took root only centuries ago in the wake of colonization and the slave trade, which were foundational in the rise of the global capitalist order.

Shane has done a great job in previous interviews explaining that side of the subject, and in this interview in particular we spend time examining the fragility of white identity. The idea that white people are under threat—where does this fear of white erasure come from, as is expressed in the manifestos of so many of these white supremacist shooters? Shane has a really deep understanding of far-right organizing and the ideological core of fascist politics, and so as societies undergo more and more stress as a result of ecological degradation, climate disruption, economic collapse, and social disintegration, it should not surprise us that far-right violence is rising as a response to unfolding trends.

As much as this is a material crisis, this also a crisis of identity, one that has emerged out of the impacts of hundreds of years of colonization. Colonization is complex and cuts both ways.

PATRICK FARNSWORTH: Anybody who looks at how European societies migrated and interacted with one another, and the kind of intercultural communication that existed between different religions, different belief systems, different ethnic groups see this: there was a lot of crossing. There was no provincial little white race society that existed at any point in human history.

Yet we tend to project these ideas onto the past. A really thick ideology bleeds out of this: "white people have always been great, why doesn't anyone see that? And we have to protect that at any cost to protect the white race from being infiltrated by these aliens, by these other people."

THERE IS AN INCREDIBLE ANXIETY THAT GROWS IN PEOPLE FROM THEIR DISCONTENT WITH THEIR LIVES.

To me, white identity feels so fragile. Anytime it feels threatened, all these acts of violence emerge out of it. I don't know if there's a question there for you, but it just gets me going because I think about the premise upon which it's built and it's so faulty.

SHANE BURLEY: Yeah. I think something here is a real internalization of a settler colonialist myth about what whiteness is and about what whiteness has meant. This idea that white people are disappearing takes a very, very narrow view of itself. For example, the idea that white people are disappearing is really an issue of skin color, the actual pigmentation of it. There's gonna be fewer people with a certain level of lightness in their skin. We all know skin comes at a gradient. I don't know at what point it becomes white, or at what point it stops being white. But when they say "the decline of white people," it really is about a number of people with a very specific phenotype.

The problem is that the genes themselves that go into whatever they believe construct these people don't disappear. They progress, they build new communities. They go on, names change, peoples change. That's just the story of human beings. Something totally artificial (it has to be socially constructed), tribal identity, is latched onto with such vigor as if it's a scientific fact. And it determines whether or not we have successful lives and will have futures. It's a really bizarre notion.

We continue to prop up a history that claims Europe's history was monoracial, that the Europeans had some legitimacy to land and things like that; as long as we continue to make that the central narrative, we're going to continue to allow an identity that's wholly toxic to foster and build up. And there's really nothing positive about this core white identity that exists.

PATRICK FARNSWORTH: In your piece with Alexander Reed Ross, you mentioned that what happened in New Zealand and what happens now all over the world—white supremacists committing acts of terror—is an extreme expression of a sentiment that is normally felt in nonextreme environments. That's a part of the problem to some degree. But this general white identity that white people feel—whether they consciously analyze it or not—is prevalent all over the United States. I think it definitely is present within many of the fans of Donald Trump especially. White anxiety seems to create an environment that allows these types of actions to emerge in the first place.

SHANE BURLEY: Yeah, I think that identity is what is being grappled with, because it's what they have. You mentioned earlier that we're being in a point of late capitalism, and I think it's really important to think about what that means. We're talking about climate collapse, total collapse, and talking about an economic system that is literally breaking apart at the seams, even while people live reasonably comfortable lives in a lot of global Northern first world countries.

There is an incredible anxiety that grows in people from their discontent with their lives. People work shitty jobs, have shitty relationships, don't have strong communities, all those things; and at the same time, they don't have any guarantees about who they're going to be in the future. Whiteness is there, whiteness is a language they're fluent in. It's really hard to then present a revolutionary alternative and say that whiteness is holding you back and is actually a toxic falsehood.

We're seeing a shift in politics away from what is sometimes called soft power—the power of negotiation, values influencing decisions and negotiation—to a place of really hard power where people, political parties, political actors, and so on battle for hegemony. We're seeing the rise of a far right that is sometimes fully white nationalist and sometimes more of an undercurrent of nativism. That's here to stay, because it's feeding on a general change in society away from a perceived stability. That anxiety only grows. I think that people who have a left alternative to that, or a revolutionary alternative, have a lot to offer in that realm. But the contention with white supremacy is massive, and it's not as though I think there's an easy solution to confront those sorts of things either.

SINIŠA MALEŠEVIĆ

There is another layer that intersects with this examination of identity, specifically nationalism and the rise of the nation-state in the modern era. In my interview with sociologist Siniša Malešević, we discuss his research into the historical formation of national identity (nationalism) in the modern era, and how the molding of national identities by nation-states over the past several centuries has given rise to unprecedented large-scale violence previously unseen throughout human history.

Siniša explains the social mechanisms and techniques the nation-state uses in order to wed the individual (the citizen) to the ideologies and interests of the nation, through a combination of sophisticated propaganda, state-mandated education (indoctrination), and warfare. Siniša expounds on the role nationalism (and its corollary, patriotism) has played in the "rise of organized brutality," a function implicit in the formation and legitimization of the nation-state.

PATRICK FARNSWORTH: I remember once hearing, and I don't know how factual this is, that when you have soldiers in a battlefield, and they're supposed to be shooting at the enemy, a lot of times those soldiers will shoot above the heads of the people that they're supposed to be aiming at. I think we have, for good reason, a natural distaste for killing other people. But there are ways in which people are manipulated and coerced, because I think ideology tricks us into thinking that we're autonomous individuals serving something bigger than ourselves. So we're willing to kill or enact violence on other people to serve the ends of that ideology, to serve the purpose of the ideology.

We have this idea that we've had nations for thousands and thousands of years, but the nation is a fairly recent manifestation. Where did this idea of the nation come from? What are some other functions of a nation that are unique compared to previous forms of complex society?

SINIŠA MALEŠEVIĆ: That's a good point. States have been around for a long time, but in a very different form. We start off as hunter-gatherers (well mostly gatherers, since hunting develops quite late), and then we have intermediary forms, not full states yet, chiefdoms where an interesting element of a complex hunter-gathering society becomes to some extent sedentary. And then we have city-states, and then these so-called capstone empires, empires which expand gradually. Then, within the European context, these various composite kingdoms or patrimonial kingdoms arise, feudal arrangements, if you like, and then nation-states develop very late.

It's only really in the last 200 or 250 years that nation-states gradually take over. We have an image of the French and American Revolutions as a pinnacle of nation-state formation, but that's really only the beginning of this process. France and the United States were still not nation-states as we understand nation-states today. It took a long time for a lot of people to start identifying with this concept of a nation-state.

Before nation-states, the much more powerful forms of territorial organization were empires. Empires were there for at least five thousand years, and they were much bigger and very different. When we focus on differences between empires and nation-states, we see that empires didn't have borders, they had frontiers. When you talk about early empires—the capstone empires—their vision of the world was "this is it." The Roman empire and the Chinese empires saw themselves as the world. Everything else was beyond civilization, and they didn't recognize other empires.

It's only in the nineteenth century that you see a recognition that the French empire is here and the British empire is there, and so on and so forth. They were not fixed entities. Nation-states have fixed borders, even though they might have an ambition to expand them; regardless their borders are fixed. Also, the concept of hierarchy is very different, because in empires social hierarchies are inherent. If you're a born aristocrat, you stay an aristocrat. As a peasant, which is 95 percent of the population, you can never become an aristocrat.

This is different in nation-states. Nation-states allow for a sense of sovereignty that comes from the people, at least nominally, so there is a sense of moral equality. That obviously does not mean that nation-states are egalitarian, it's just that the principle of moral equality is an important element of legitimacy within the nation-state.

Another difference is that empires are universalist, claiming the whole world in the name of religion or a civilizing mission or some broader ideological approach. Nation-states have a definition and are particularist: they justify their existence through some form of nationalism, invoking the idea that it's our people, we represent this particular people.

These are some of the key features. But the nation-state is modern. They have been around for 200 to 250 years, but it's really after the Second World War that you see the nation-state become dominant. It becomes a hegemonic form throughout the world, and they are not contested anymore. So even cities, small city-states like Singapore or Liechtenstein, now frame themselves as a nation-state, even though they don't really fit the traditional typologies. This is because the nation-state principle has become so dominant, so ingrained, that it is seen as the only legitimate way.

Japan is the only one which is officially empire, but in reality it isn't. Empire has become a very negative term, which wasn't the case in the eighteenth or nineteenth century. It previously had a positive connotation, which means the nation-states have captured the ideological power that empires and kingdoms had. That is a dramatic change.

What's also important, if you look internally within the nation-states, is the sense of how we see it, or how the majority of people see themselves. They see themselves in national terms. Nation is somehow assumed to be normal: it has become naturalized. We are all born into a nation-state. You can move from one nation-state to another. You can get citizenship from France, Germany, or wherever, but there is no escaping the world of nation-states unless you live somewhere in the middle of the Pacific or the polar caps. The world is completely now carved out by the nation-states.

We are born in nation-states, and we live through institutions of nation-states. They shape our daily life and our entire life. They've become deeply embedded. I call them "grounded nationalisms," because they've penetrated every aspect of social life. It's very difficult now to escape the nation-state. There is an assumption that this is a natural way that the world should be divided, but it isn't natural. It's very recent.

PATRICK FARNSWORTH: Sometimes I feel like an alien when I see people pledging allegiance to a flag here the United States. Nationalism has always been present, but it's been especially present in the past several years. I think as nations enter into states of crisis, nationalism seems to increase. This nationalistic fervor seems to really increase, and I've always had very mixed feelings about it. I've always felt like that it's a very strange notion that people attach their own personal, individual identities to the nation. "I am an American," or "you are Irish," or "you are this..."

We have created our personal identities with this thing called the nation as if it's old, but that doesn't actually seem to be the case at all. Obviously people are often connected to their tribe, or some larger organization or group, but we seem to have reached a new level of abstraction.

SINIŠA MALEŠEVIĆ: Sure. I want to say how we've come to where we are. It's a long-term historical process, an ongoing process, and we tend to see it more when we have these crises like Brexit and Trump and things like that. Nationalism becomes much more visible, and then we are shocked. Intellectuals in particular seem to be shocked every few decades. They see this "crazy" nationalism. But nationalism is much stronger when it's less visible.

This is connected to the power of the organizational capacity of the nation-state. After the French Revolution, and in the mid-nineteenth century, very few and mostly middle class people identified with the nation. There is a very good book by Eugene Weber, *Peasants into Frenchmen*. He shows that even in France—which was the epitome of the nationalizing, centralizing state—the majority of people didn't see themselves as French as late as the nineteenth century or early twentieth century. It was still a middle class phenomenon at that point.

People in Paris did see themselves as French, but people in other regions, in the villages, still identified more in local terms: in terms of kinship, clan, and religion rather than in terms of the nation. Nationhood as a fully fledged sociological fact develops very late. It develops because of institutions of the state's educational system: high literacy rates are very important, because if you don't have a literate population it's very difficult to penetrate them ideologically. But if you put them into schools and teach them history in a particular way, and if you teach them a national language which becomes standardized and then imposed, the work is easier. This is a reason why primary education and now secondary education in many countries is compulsory. This is not only about teaching certain skills, it is about making people national, and that's what the educational systems have done.

In the late nineteenth century, after the Franco-Prussian war, when France lost they devised these subjects that we all now have: national history, and geography, and language: that was imposed deliberately, to make people into loyal French subjects. The Pledge of Allegiance that you mentioned is also one of these mechanisms that make young people into loyal Americans, or loyal French, or German or Indian or whatever.

Nationalism has become powerful precisely because of these nation-state mechanisms, because the nation-state is seen as the only legitimate form. Whoever is in power frames individuals along these lines. Obviously we see Trump as this excessive individual, but during the Obama years the Pledge of Allegiance was still there, the inauguration of the president was still there, the nation-centric language was still there. So now we see a much more aggressive version of the same thing. This is not a break with the past, it's just a little more of a caricature of what already exists.

It's really important to recognize that the level and scale of violence human beings are capable of emerged quite recently, as Siniša has explored in his work with great clarity and depth. Not all nationalisms are bad, but indeed, any form of nationalism that emerges through oppressive, systemic violence must be resisted and countered at every turn. This form of national identity divides us along arbitrary lines and presents the possibility of mass violence.

The human cost from the establishment of nation-states over the past few centuries has been horrific. As with all things on this scale, it's complex, but when the subject is brought down to the human level, the nature of nationalism is often that of exclusion, dehumanization of "the other," and the traumatic and lasting impacts of war.

NATIONALISM HAS BECOME POWERFUL PRECISELY BECAUSE OF THESE NATION-STATE MECHANISMS, BECAUSE THE NATION-STATE IS SEEN AS THE ONLY LEGITIMATE FORM. WHOEVER IS IN POWER FRAMES INDIVIDUALS ALONG THESE LINES.

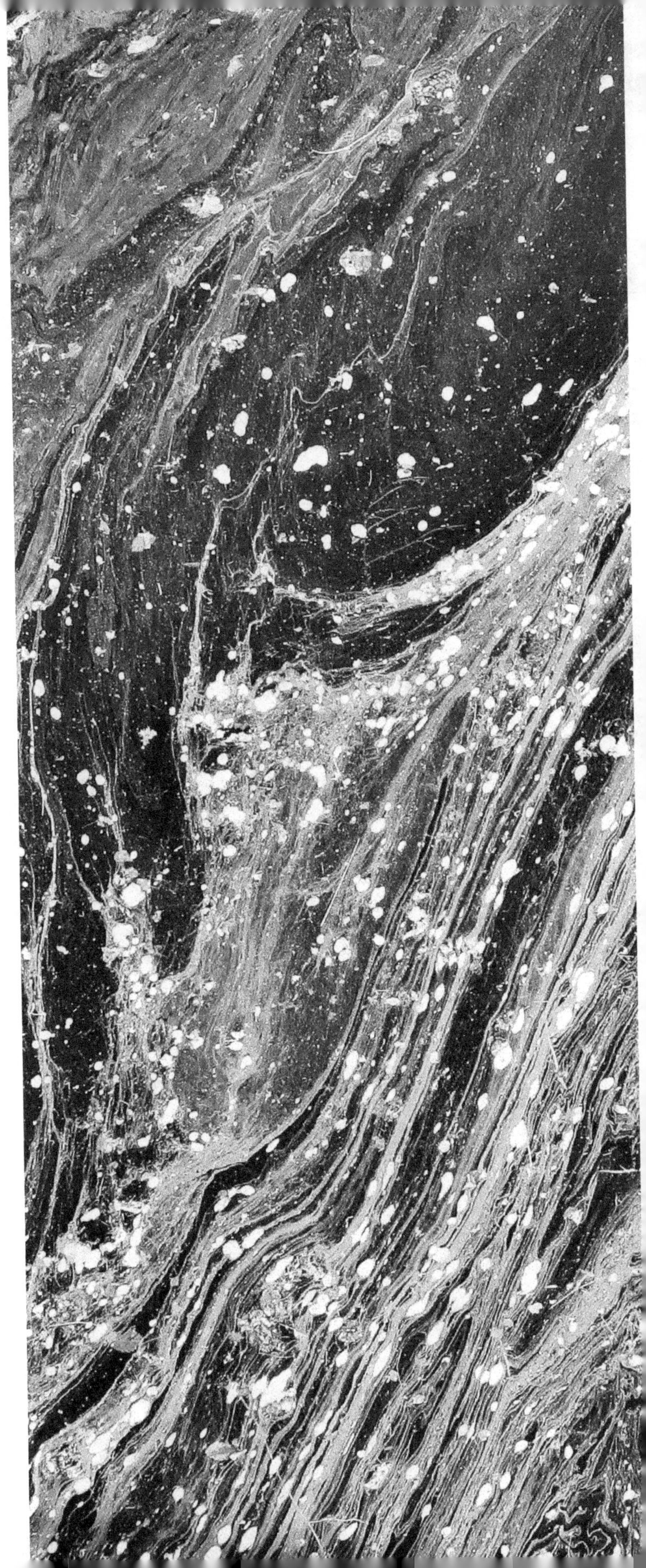

LIYAH BABAYAN (WITH CYNTHIA JONES)

In 2018, I had the great fortune to work with Cynthia Jones, artistic director of the modern dance company Inspirata Dance Project, based in southern Idaho. We collaborated on three interviews with refugees in the local community, which I released as podcast episodes, with audio segments of those discussions used in the final dance production. The event was titled *Koinonia* and was presented at the College of Southern Idaho in May 2019. The first interview Cynthia and I conducted was with Samra Culum, the second with Milica Popović, and the third and last interview with Liyah Babayan. The interview I would like to present is with Liyah. I've interviewed her three times in total, and this was our second. In fact, the very first proper interview I ever conducted for the podcast was with her.

Liyah recounts her childhood experiences of fleeing the pogroms enacted against the Armenian minority population in Baku, Azerbaijan in the midst the collapse of the Soviet Union in 1989–1990. After experiencing incomprehensible trauma and dehumanization, Liyah and her family fled to Armenia, where they were homeless for over three years. Liyah recalls the hostility and derision (with notable punctuations of deep generosity) she and her family experienced from her fellow citizens during this time, as is too often the case with displaced and traumatized refugee populations around the world.

After this period, Liyah's family was finally granted the refugee status required to make their way to the United States, and they ultimately resettled in Twin Falls, Idaho through the College of Southern Idaho Refugee

Resettlement Program. While Liyah and her family were fortunate enough to be able to escape the horrific violence in Baku and their desperate living situation thereafter, their journey toward integration and healing was only just beginning.

Listening to Samra, Milica, and Liyah express their life stories through this artistic collaboration with Cynthia opened me up to something really essential in the human condition. Those interviews were some of the most profound conversations I've had the honor to be a part of.

LIYAH BABAYAN: When you're hunted as a human being, when you feel hunted like an animal, some of the emotion you have isn't human-like. You go back to this most primitive feeling. It's not human. I'm not talking about the fear of the dark and the fear of a bogey man and the fear of losing your family: those are pretty common fears. But when you're hunted, when you feel hunted—for whatever reason, it doesn't even matter if it's religious or ethnic or whatever—it all comes down to that same primitive animal low frequency, and those emotions? They're not human. They're not. They're not supposed to be part of the human experience. That's why we are civilized.

And there's no room for those emotions in a civilized setting. That's why a lot of people will go silent, or go within and internalize their trauma and their experience. There is no language for it, and most of it is so unbelievable that it is difficult to express. So yeah, that base—that low. We pride ourselves as a society that we get to this point where our little children are supposed to feel this way, and anything below that is a taint on their childhood and worthy of coddling, worthy of treatment, and of nursing and of healing.

You know, there's a standard, and in every society that standard is different. You go to some societies and that standard is way down here, like six-year-olds are working and hunting and providing, and they don't get therapy for that. But when society collapses and when there's war, when we're barbaric and there's ethnic killing, that whole standard of emotional development—especially for children, children of war and children of these conflicts—that just goes away. There is no low anymore.

The lowest existence for humanity is when society dissolves, and order and safety disappear. That happens to a lot of people who are escaping atrocities, and like they're being hunted, they're feeling that threat on their life, and it's a pure threat from another human being on your life and not because you committed a crime. It's not because of something you've done. There are not any kind of rules to go by in that.

There are very strong feelings that are very difficult to express when someone goes through this, that feeling of being disposable. Feeling your humanity disappearing, becoming subhuman and disposable—those don't just go away when you're safe. You don't just land here and the Statue of Liberty waves at you. You are not all of a sudden upgraded to Human 2.0 or whatever.

PATRICK FARNSWORTH: Like all of your experiences disappear.

LIYAH BABAYAN: Yeah. Feeling subhuman stays with you. That is your trauma. Being regarded as disposable, the lowest of the lowest when it comes to human life, even more disposable than material objects. There's art that's more valued than human life! Yeah. There's that Jesus cheese thing, like that cheese that...

PATRICK FARNSWORTH: ...has Jesus, his face on it.

LIYAH BABAYAN: Yeah! And no disrespect to Jesus. Jesus is my friend. But what I'm saying is there are pieces of cheese that are valued more than human life.

PATRICK FARNSWORTH: Yeah.

LIYAH BABAYAN: People will go into the streets and say "save that cheese."

When your psyche comes to grips with this.... Imagine this, like I'm a kid that is just innocent. I'm just feeling things, like "oh, that does that when I touch it, and that dandelion is magic." When you're a kid you don't think dandelions are weeds, they're magic, they're like wands and your wishes are supposed to come true when you blow on those dandelions. So you're a kid, and you're just sensing the world, and you're piecing together what you can even remember day to day, because you go to sleep and you wake up and only stored a tiny bit of your experience because you're a kid!

And you're eating, you're playing, and you're not thinking about it—you're just in the process of just being there, and being a kid. Then all of a sudden you're shown (you're not told but you're shown) that you're not worthy of life. Your humanity is negotiable.

Maybe you don't understand your own humanity, but you can see that these adult people in your life are shaken. And these people are pillars to you. They're unshakeable—your grandparents and your parents, if they look at you a certain way they're gods when you're a kid. So when you see the adults shaken, and when you see that they're becoming smaller and dimmer, fearful and hiding and minimizing themselves, minimizing their humanity, minimizing their wholeness: you learn naturally that well, where were those people? We're disposable, we're the disposables. No one has to tell you. You see that because you become homeless, and no one cares, and your family members are slaughtered. And that's normal. That becomes normalized.

Liyah gives us a glimpse into what it's like to be on the other end of this, to be the one that's targeted, to be the one that is othered and forced to bear an almost unbearable trauma.

If we're going to build something beautiful in this time, we have to do everything we can to counter fascism and authoritarianism, and the first step in that resistance is to counter dehumanizing narratives. We have to establish ways of living and being in this world that encourage solidarity and understanding between people.

It's easier said than done. I'm not necessarily someone that knows even how to do this, but there are plenty of people who do, or at the very least, can indicate what's possible in this time.

THE LOWEST EXISTENCE FOR HUMANITY IS WHEN SOCIETY DISSOLVES, AND ORDER AND SAFETY DISAPPEAR.

PART FOUR
BEAUTIFUL RESISTANCE

CONTEXT IS IMPORTANT when we ask the question "what is human nature?" The question itself may be limited to ideological frameworks of which we are barely conscious, as demonstrated by the previous explorations into the historical processes that have produced these crises.

We are capable of cognitive blindness, especially within the context of the mass society. We perpetuate artificial differences and separations between ourselves and others, whether that be other human beings or nonhuman life. Often, this generates patterns of behavior which serve the ruling classes, and these conflicts and divisions are often framed and stoked by them. That's how it has worked for much of recorded human history.

Sometimes when we look at disaster situations, when the social order collapses (as often happens with complex large scale societies), it's assumed that people are just going to start killing and stealing from each other, that unfettered barbarity will ensue. But this is not always, or even often, the case. Often these situations play out differently, and there are organizational forms which encourage another side of human nature: the side of solidarity and cooperation. This is especially important to keep in mind now, because building resiliency through networks of decentralized forms of cooperation and solidarity will be key to managing the impacts of infrastructural collapse.

DEZERAY LYN

To amplify this perspective, I would like to present my interview with Dezeray Lyn. Dezeray is an activist, street medic, and co-organizer with Mutual Aid Disaster Relief (MADR).

Mutual Aid Disaster Relief is a national network made up of many eco-activists, social justice activists, global justice activists, permaculturalists, community organizers, and others who are actively supporting disaster survivors in a spirit of mutual aid and solidarity.

Instead of relying on a nongovernmental organization (NGO) or on a government agency like the Federal Emergency Management Agency (FEMA) to provide all the supplies and support required in a community's time of need in the midst of a natural disaster, MADR works directly with the community and provides grassroots support. Volunteers demonstrate that centralized institutions, like government agencies and nonprofits, are not only inefficient but insufficient in addressing the immediate needs of the impacted community. Grassroots decision-making and an intense focus on respecting the autonomy of the community is made the priority, and the results of this approach are astounding.

In times of collapse and disaster, communities often automatically begin to function in this way, when the impositions of the state aren't a factor. It sounds contradictory, but when the veneer of civilization and organized society recedes, people are capable of powerful things. One thing they are certainly capable of is communally aiding one another, to work within the community to find the resources required to provide some sort of sustenance, stability, and social cohesion. That is what MADR has demonstrated and enhanced in numerous communities around the United States and abroad.

In my interview with Dezeray, I wanted to convey what is possible not only in times of collapse, but what has always been possible if only we can think about what we want in this world. What is the most beautiful way that we can conduct ourselves in our climate disrupted present and future?

PATRICK FARNSWORTH: I really like that you're amplifying something that people are not only capable of doing but have almost an instinctual drive to do anyway. When a disaster strikes in a community, when things really do fall apart, there is an idea that people start shoplifting and looting, and chaos ensues and people start fighting. It's a myth that we're told.

You've been to Puerto Rico after Hurricane Maria hit. You worked on the ground there. What is your experience with how people react to these disaster scenarios as climate catastrophe continues to unfold on this planet?

DEZERAY LYN: You mean the impacted communities themselves?

PATRICK FARNSWORTH: Yes. How do people react to these disasters, in your experience?

DEZERAY LYN: In every situation that I have been a part of, I've witnessed communities doing the work to take care of each other. The biggest hurdle that communities have faced in my view is barriers and bottlenecks that these government institutions and nonprofits have put up for them. But there was so much autonomous community response, particularly in Puerto Rico with these centers of mutual aid that opened up across the island.

The government completely failed to give any kind of response. Actually, it even criticized the people as if they had no interest in their own recovery, as if they were doing nothing to pull themselves together. This was exactly the narrative that was propped up again and again. Yet the whole time this narrative was being circulated, the people of Puerto Rico had pulled together: they had created community kitchens with no power, with no running water, with their streets in ruins, with their homes in ruins, with their roofs gone. They were offering hundreds of meals a day in the community. They were organizing autonomous water brigades and spaces where people could get medical support and medical access. People were organizing for their neighbors to be able to get access to things like insulin.

It continues to this day, a really powerful movement from the streets, from the grassroots. It's a stunning movement, and it continues to support people in the rebuilding process. I've seen it again and again and again: in New Orleans and the Lower Ninth Ward, in the Carolinas, across Florida. It's been something that repeats itself, because people revert to a natural state of cooperation and autonomous response, even in the face of the states setting up these barriers, criminalizing people for this "looting," which is another toxic narrative that's put out because capital needs to defend capital. That's its first priority, and its only priority, at the expense of people getting things they need to survive. In the Carolinas, where a dollar store was set up in a Macy's or a Saks Fifth Avenue, the manager was giving people things. The police moved in and arrested them, charged people that had undergone a severe and deadly hurricane. They were charged with felonies, and the government changed looting to a felony offense.

IN EVERY SITUATION THAT I HAVE BEEN A PART OF, I'VE WITNESSED COMMUNITIES DOING THE WORK TO TAKE CARE OF EACH OTHER.

Really, the cooperation I've seen in these communities has been really incredible.

PATRICK FARNSWORTH: It's amazing how that side of it is suppressed so thoroughly. There is this sense that we need to send law enforcement in to keep order, but people are inclined to keep order anyway. We have this Hobbesian world view, that if the state doesn't come in and suppress our inherently violent nature, then we're just going to start looting and killing each other. But as you point out, communities will start to support each other, try to feed each other, and will support each other.

How does the organization of mutual aid disaster relief work exactly? How are resources distributed? How do different networks communicate with each other and understand how to coordinate activity? How does all of this play out in a real-life scenario?

DEZERAY LYN: So, since this is decentralized and it's survivor-centric, it looks different every time. Different communities and different people work in different ways, and it's organized, populated, and engined by people.

It's different folks involved in each disaster in those communities, and people also come from offsite. There is always offsite support coming in to provide assistance, to share skills, and to help out. We have things like wish lists, where people determine what they need for that community: different tools, different food items, different mold remediation gear, and other things like this. They decide what they need, and we amplify those needs through sharing wish lists and sending any support we can their way.

If we can, we organize offsite teams that can go there and respond to these disasters, and we listen to the communities that are already organizing on the ground. We further populate the movement, work with them to help support and provide access to things like medical aid and tools. When funds do come in, the funds go to an all-volunteer network. We provide support for transportation, bringing in things like fuel that are not readily available on the ground. The funds go for things like medical solidarity, creating clinics, for getting box trucks into the area. In Panama City right now, there's a crew that's been there just a few days after Hurricane Michael came through. There's a clinic, there's a tool lending library. Every single day there's mobile distribution, trucks, cars, and caravans bringing things to folks in impacted communities.

To talk about one unique issue in this particular area of Panama City that we've done: there's been a lot of predation and opportunism by the state, so we've done a lot of work on tenants rights. Landlords have come into the projects that were damaged—some damaged only a little—and illegally evicted all of the residents, some with an eviction notice that was a five sentence typed page.

These tenants? There are families that have people who are amputees, or people with severe medical issues. They're out of work because their work is heavily damaged. They have limited access to fuel. And the landlords are telling them in bold lettering: leave immediately. They're not given contingencies or being told where they can go, and the landlords make no consideration for the fact that people need to stay in that area because their doctors, their children's schools, their family, and their support networks are there.

In every community it looks different, but really our work is just amplifying the funds that come in. The work and decisions are community driven. No one that's not on the ground is making decisions about what specific funds are allocated for, of where supplies are going. It's the autonomous movements on the ground that we really support, we join with them and support their self-determined needs.

PATRICK FARNSWORTH: How did this idea come into existence? I know that mutual aid is as old as humanity itself, but this specific network: where did it come from? What was the beginnings of this organization?

DEZERAY LYN: I and a lot of the folks involved see the founding in the autonomous response in the Lower Ninth and other areas around New Orleans after Hurricane Katrina. Then, there were folks that did work in Haiti after the earthquakes, folks that did work after Superstorm Sandy with Occupy Sandy. Folks had been involved in all these other mutual aid networks, then tried to make something that wasn't so disaster- and region-specific.

Occupy Sandy was very specific for that disaster, as were efforts in New Orleans. And these disasters are increasing. Literally every storm you hear about has this attached adjective, that it's "historic, it's historic, it's historic." These are just increasing.

This mutual aid network started a couple of years ago as a non-disaster-specific, non-region-specific network. It's just like Food Not Bombs. It's a movement that can have autonomous pockets and autonomous, decentralized movements everywhere. But it goes on the same guiding principles, the same vision of empowerment, of solidarity-not-charity, and building power from the streets.

This is done with the people in the impacted community themselves, rather than—and vocally and visibly against—the actions of federal government occupations like FEMA and the resources they subsume in the face of this. They create bottlenecks, issues, tensions, traumas, and anxieties, and the atmosphere they stir up by their presence. They come in with a strong bureaucratic hand and all of this red tape, but that's the last thing people need. People need quick access.

So, a couple of years ago, Mutual Aid Disaster Relief was founded, mainly with folks who were doing autonomous work for over a decade since Katrina. Since then, sadly, there have been yearly increases in these disasters. There are many intersecting issues, as now with the fires in California, such as inmate justice, because inmates are out fighting fires for a dollar an hour and getting injured. This matter intersects with a lot of other movements.

MUTUAL AID ISN'T JUST THE NAME OF THE GROUP. THIS IS A WAY OF BEING, IT'S SOLIDARITY NOT CHARITY.

I think in the last two years, the work has been opening doors, opening pathways, removing bottlenecks, and

trying to connect people with each other so that support networks are available for communities that endure these climate catastrophes.

PATRICK FARNSWORTH: Doing this on the ground level in a decentralized fashion: in a political context, that's anarchism in practice. That relates to this idea of mutual aid being something that's very natural in the human experience.

One of the most famous anarchists, the Russian anarchist Peter Kropotkin, wrote *Mutual Aid: A Factor of Evolution*, which was the idea that people don't thrive or develop fully into human beings under hierarchical oppressive institutions and top-down forms of social organizing.

I think this network you are a part of embodies these notions, and I think people can understand why, when a disaster strikes, the best way to approach these situations is through this anarchist principle. But I think people also have a false understanding of what it means to be an anarchist. Could you comment on the principles of mutual aid and the political framework of these ideas?

DEZERAY LYN: I wouldn't attach a political name to this work, because there are folks from different political ideologies that are involved, and this is autonomous. It's reverting to an actual state of being: cooperation. But mutual aid isn't just the name of the group. This is a way of being, it's solidarity not charity.

We believe that, rather than treating survivors of climate catastrophe as powerless consumers, or rather than just handing them things and then leaving, we're committed to understanding, respecting, amplifying, supporting, and backing people. The fact is, people have a stake in their own survival and in the rebuilding

of their communities. They know their communities, so solidarity not charity is really important.

This principle is also really important for exercising dual power, which is creating the world that we want to see, while challenging and resisting the one that is currently wreaking havoc and chaos on people who are trying to struggle through it. Solidarity not charity is a major tenet of mutual aid, as well as being vocal against forces that subsume resources and create bottlenecks. Another major tenet is being survivor-centric: listening to communities and letting those autonomous movements be the ones to lead the efforts.

Our work is to support those efforts, moving with them horizontally, and working on consensus-based decision-making processes. When you're acting with mutual aid in disaster situations, there is no place for hierarchical structures. And in a larger sense, capitalism itself is a disaster, as are the things we have to endure: mass incarceration, concentrated poverty, environmental racism, and all these other issues. These all come about because of capitalism.

As Dezeray explains, it's not only about aiding people, it's also about encouraging the flourishing of something that comes innately to us as human beings. When it comes to these disaster scenarios, we must note that the various institutions, governmental or otherwise, that wish to impose their idea of "aid" on disaster-impacted communities are really just trying to reassert their position in the social order. It's less about aiding the community than it is about imposing an order that benefits them, at any cost.

PETER GELDERLOOS

What are the actual tactics that we need, how do we hold places of power to create networks of solidarity? How can we resist and upend the systems of power that have produced the climate crisis to begin with?

Drawing upon the history of effective resistance movements and their tactics can help in this regard. To explore this further, I'd like to present a part of my interview with anarchist, activist, and writer Peter Gelderloos. Peter is the author of *How Nonviolence Protects the State*, which examines the most lauded "nonviolent" movements in recent history.

Peter examines two specific examples of resistance: the anticolonial movement in India in the early twentieth century (in which Mahatma Gandhi became well-known for his use of nonviolent resistance against British rule), and the civil rights and antiwar movements in the United States in the 1950s through the 1970s. As Peter elaborates, nonviolence and pacifism severely limit resistance movements in upending entrenched systems of oppression and violence, of which the State is the ultimate manifestation.

PATRICK FARNSWORTH: Your book points to the anticolonial movement in India with Gandhi and the civil rights movement here in the United States from 1950s, 1960s into the 1970s, as well as the antiwar movement against the Vietnam War, and you lay out how these movements have been whitewashed. Can you talk more about this?

PETER GELDERLOOS: The book is not an apology or a call for violence. The book is a criticism of nonviolence, and a part of that is the criticism of the entire dichotomy between violence and nonviolence. In all of these historical examples that you just brought up, it wasn't one action that brought about the change. It was a complex interrelationship of many different tactics. Unless we realize that it is uncontrollable, complex, and very heterogeneous movements that bring about profound change, I don't think we'd be able to act intelligently within movements of social conflicts.

With the case of India, it's no coincidence that Gandhi is the only person that we hear about in the history textbooks, that Gandhi is so famous. George Orwell said that the winners write the history books. And at the same time, proponents of nonviolence will say nonviolence in such a dangerous idea that the elites don't want us to learn about nonviolence. This is so patently absurd, because in state-sponsored education, in corporate media, we're again and again told how important Gandhi was.

In the actual history, though, there were a lot of other formations equally or more important than Gandhi—for example, the Indian National Congress, which for much of its struggle tended not to favor Gandhian methods. There was also an entire network of armed anticapitalists in the late nineteenth century and early twentieth century in India. There was an international movement of Indian exiles who were very close to anarchism, who also took part in anarchist movements elsewhere while fighting for independence for India from British colonialism, like the Ghadar Party. This is something dealt with more extensively in Maia Ramnath's *Decolonizing Anarchism*, a very interesting book.

THE BOOK IS A CRITICISM OF NONVIOLENCE, AND A PART OF THAT IS THE CRITICISM OF THE ENTIRE DICHOTOMY BETWEEN VIOLENCE AND NONVIOLENCE.

There were moments, obviously, when Gandhi's protests were extremely important. But there were other moments when major riots were important. There were moments when these huge, armed structures were more important. And there was never a need for internecine competition between those different divisions, with one exception, which I'm going to clarify in a moment.

There's this breach of solidarity that stems from the authoritarian impulse within nonviolence: they had to claim absolute credit for any change that happens. They have to prohibit other people from using other tactics, and so they systematically turn on other people within the movement. That's something that Gandhi did repeatedly. At times, you also see a great deal of support in India for canceling protest movements after confrontations that he didn't approve of cropped up.

I mentioned that there is an exception to this idea that we can use different tactics, we can use different methods and still be in solidarity with one another. Those of us who might advocate sabotage or self-defense or more combative tactics have absolutely no need to prohibit peaceful marches or acts of civil disobedience. We have no need to talk badly to the people who do those things, the way that they systematically deal with the actions they don't approve of.

The exception to this is when you have a partisan movement that is authoritarian, that is trying to seize control. You can see authoritarian currents within the anticolonial movement in India and in other movements in the world that are only trying to take over state power, to put themselves in charge to reap the benefits of struggle. You also have antiauthoritarian currents which are trying to decentralize power, so that power can become a collective process in which everyone takes part in shaping your own life, shaping the world around them rather than living according to somebody's plan of a better government or a better society.

So really, the only moment in which there isn't a possibility for solidarity, despite difference for heterogeneous struggles, is when a current of the struggle doesn't respect the basic legitimacy of other currents in that struggle and tries to dominate the entire struggle. Authoritarianism is the only element that makes that kind of heterogeneous solidarity impossible. Authoritarianism can certainly be found in a number of armed or otherwise combative movements throughout history. But I argue that it's a systematic part of most iterations of nonviolence throughout the twentieth and twenty-first centuries.

PATRICK FARNSWORTH: It seems, as with Gandhi, some try to co-opt the movement and to distance themselves from what they consider to be the more violent elements of a movement. But as soon as things move in their favor, they tend to take control of the power structure. They become the new state, and I think we need to remind ourselves that's a thing that happens over and over again.

I was thinking about people who don't want to engage in what they consider to be violent behavior when it comes to protest, and want to distance themselves from that. But do you think there's often within these more nonviolent segments a more cynical element to that? Because I actually have sympathy for people who say, "I'm a pacifist. I don't want to hurt anybody, not even the people oppressing me. I just want to use nonviolent tactics to get something done here."

It seems like the people you're talking about are willing to take advantage of the momentum generated by the diversity of tactics in order to gain power and control over others. Is that something you see implicit in nonviolence itself, or is that something you see within particular groups that use nonviolence as a platform to get into positions of power?

PETER GELDERLOOS: I think there are a great number of people within nonviolent movements who are committed to making the world a better place. They're committed to whatever cause they're a part of, they're sincere. Some of them have a personal preference to avoid certain types of situations or certain types of conflicts, and they try to extend that as practice. Others have a certain ideology of "do no harm" and they're coming from a more radical Christian background or similar. I see nothing wrong with that. I've had a number of friends from this movement.

But those movements also tend to provide useful platforms for what we could refer to as the politicians of the movement, the ones who want to win the role of respectable intermediary, the ones who will be quoted by the media, the ones who will get to negotiate with institutions of power. Sometimes people launch their careers from nonviolent campaigns of civil disobedience and end up becoming politicians.

I would say that there's room for cynical or authoritarian people to take advantage of nonviolent platforms, without a doubt. But by no means would I say that every or even the majority of proponents of nonviolence do it for those cynical or self-interested reasons.

PATRICK FARNSWORTH: You make this claim that the civil rights movement really didn't meet its stated aims and goals. What were those goals, and why would you say that they did not meet them?

PETER GELDERLOOS: Of course, legal segregation was ended by the U.S. government in response to the civil rights movement. But if you take economic indicators, access to education, and other things like that, in a lot of ways things for black communities have gotten worse.

If we talk to the veterans of that movement, if we read the writings and the testimonies of those who are no longer alive or those who were in prison, we can see the movement was, in their own words, about a lot more than just a couple of laws on the books. It was about an entire white supremacist system and its effects on everybody's lives. That system hasn't gone away. So it seems a little bit insincere to claim victory for the civil rights movement.

Absolutely, an important battle was won. But first and foremost, it's the United States government, it's the inheritor of these white supremacists systems, that claims victory. That's a very intelligent way to win a war: make your enemy go home, make your enemy think that they've already won, especially if you're in charge of educating children. It was a very cynical maneuver to co-opt Martin Luther King and claim victory there, precisely so that people would stop fighting against the continuing forms of white supremacy.

The boycotts and the peaceful protests were an extremely important part of the civil rights movement. But that's not what won an end to legal segregation. In Birmingham, Martin Luther King was basically repeating the Albany campaign, which was a total failure. The tactics of civil disobedience didn't work there. All of the organizers ended up in jail without changing anything. The difference is that in Birmingham, the local youth started fighting back against police. They didn't consider it dignified to just let themselves get beat up and jailed. And they didn't find it dignified to listen to these college educated, out-of-town professional activists who were telling them how they should resist.

So they rose up, they fought back. They kicked the cops out of the center of Birmingham. They burned all of the white businesses. And just a few weeks later, Birmingham, which was the bastion of segregation in the South, which was supposedly never going to desegregate, started desegregating. Then, the executive of the U.S. government began supporting the Civil Rights Act, even though before he kept telling Martin Luther King that now wasn't the time. So it was also the fear of the reality of combative collective self-defense by black communities in the U.S., and the very real and growing possibility of black uprisings in the U.S., that accomplished that partial victory.

PATRICK FARNSWORTH: I remember when the Occupy movement was happening, and there were real debates about this, that sitting in a park all day and trying to not get arrested by the police isn't going to really do anything. Many were saying the movement needed to evolve, to move in a different direction, to incorporate different tactics. I remember that was really when the division within the movement really became apparent, and then it started to disintegrate.

The argument that some nonviolent activists offer is that, as soon as movements become violent, we begin to see problems, more state repression, and divisions within the organizations themselves. What would you say to that?

PETER GELDERLOOS: I would say that the Occupy occupations that lasted the longest, that accomplished the most, and that engaged the most people were precisely those that used diversity of tactics and didn't constrain themselves to nonviolence. I would also say that argument reflects an extremely naive and reactionary view of the state. The state is not a passive set of institutions. It doesn't just sit there waiting for people to start misbehaving before punishing them.

All modern states use a counterinsurgency philosophy with regard to their own populations. In other words, they view their own populations as a constant threat, one they have to systematically and continuously surveil, discipline, and repress. Repression is an everyday activity.

Europe makes an interesting case study, because you have a lot of countries right next to each other which are economically similar but with very different histories in their social movements and different political strategies used by their states. The countries in Europe with the most repression, the most invasive police presence, and the most social control would be the United Kingdom, Germany, the Netherlands, and Switzerland. Some of these countries have certainly had strong social movements in the past. If you look at the years in which there is social peace, in which there is not a lot of conflict, you also see state repression advancing by leaps and bounds. On the other hand, if you look at the European countries where there are constant uprisings and constant resistance, more often than not the states are less able to develop or increase their repressive capacities, because people just don't let them.

ABSOLUTELY, AN IMPORTANT BATTLE WAS WON. BUT FIRST AND FOREMOST, IT'S THE UNITED STATES GOVERNMENT, IT'S THE INHERITOR OF THESE WHITE SUPREMACISTS SYSTEMS, THAT CLAIMS VICTORY.

States don't just wait around until there's a revolutionary movement to start being repressive.

Along with recognizing the inherent value and utility of using a diversity of tactics to resist oppressive power structures, it's also vital to unpack our assumptions of what forms of violence are acceptable and what aren't. As Peter alludes to in that segment of our discussion, there are plenty of reasons to be skeptical of the inherent "morality" of nonviolence, and the false dichotomies that attempt to present distinctions between violent versus nonviolent forms of resistance. Diversity is key here, especially when we examine what actually furthers the aims of social justice, anticapitalist, and anticolonial movements.

GORD HILL

Continuing this discussion of violence and diversity of tactics, here's part of my interview with Indigenous artist, activist, and graphic novelist Gord Hill. Gord is the author of *500 Years of Indigenous Resistance* (which was made into a graphic novel), *The Anti-Capitalist Resistance Comic Book*, and *The Antifa Comic Book*.

A major theme in our conversation is the growing intersections between what can be defined as traditionally Eurocentric leftist political theory and praxis with anticolonial Indigenous resistance throughout the so-called American continents, from North to South. Gord, having one foot in Indigenous resistance movements and the other in anticapitalist and antifascist organizing, provides needed context on how radical leftist theory and practice (for the most part, European in origin) increasingly works in tandem with anticolonial Indigenous resistance movements.

Various strains of resistance—anticolonial, ecological, anticapitalist, antifascist—can only benefit from learning and working with each other. In great part, this is already happening. Each of these respective strains of resistance often utilize a similar diversity of tactics in resistance to the State, the logic of capital, and the settler-colonial culture that dominates the land up to the present moment.

PATRICK FARNSWORTH: Before we started recording, you mentioned the Eurocentric view of anarchism, that when we talk about antifascism or anarchism, it's often through Western theory and a Western lens. But there's so much that can be learned from Indigenous struggles and Indigenous forms of organization in fighting oppressive power structures, and I think you've done a great job melding these together. In your work, in what ways have you tried to address that Western bias in anarchism?

GORD HILL: Well, anarchism is a European ideology. It arose in Europe in response to the rise of capitalism, and it was part of the socialist revolutionary movements that arose in the 1800s and early 1900s. This is a movement organized primarily by the working class to deal with this new form of economic power—capitalism. The anarchist movement (as well as the socialist and communist movement)—was born from the proletariat fighting against the rise of the capitalist state.

Some think that anarchism should also be an Indigenous thing. I don't think that's really necessary, but I do believe that it is useful to Indigenous peoples because we're colonized. We're living under the state, and anarchism is a very useful tool for understanding the modern state.

I'm involved in both Indigenous anticolonial resistance and the anarchist anticapitalist resistance. I see the importance of both, and I try to build links between them, because I think Indigenous people can learn a lot about the state from anarchism. We didn't have the state; it's not part of our tradition. We never had to deal with centralized authority and power, and so anarchism can give us a better understanding of the state, how to resist it, and how it works.

Anarchists living in North America can learn a lot from Indigenous people, because Indigenous people show another way of life is possible outside of the state. We existed for thousands of years without a state before colonization. Also, Indigenous culture, because of its connection to the natural world, can give a view of another way of life, another way of looking at the world we live in not based on capitalism, communism, or building better factories so the proletariat can run society. Indigenous culture would be contrary to that, because it's about living within the natural world, not occupying it and taking as many resources as you can.

WE'RE LIVING UNDER THE STATE, AND ANARCHISM IS A VERY USEFUL TOOL FOR UNDERSTANDING THE MODERN STATE.

So, I think both movements can learn from one another, and both are important. In Canada, Indigenous peoples are often at the forefront of anticolonial and even anticapitalist struggles, even though a lot of Indigenous people aren't anticapitalist. They're engaging in struggles that limit the ability of corporations to come in and take resources. Indigenous people in Canada are a bigger percentage of the population than in the United States, and because of the location of many Indigenous communities, Indigenous peoples in Canada can have a large impact on the politics of Canada. It's an important movement for anarchists to understand, one that will enable all of our movements to work better together once we understand each other's politics better. I think that would be a big help.

PATRICK FARNSWORTH: Do you sense that's happening?

GORD HILL: Oh, definitely. I've seen it since I've been involved, since the late 1980s. I've seen a big progression over the last few decades. In the late 1980s in Vancouver, you'd ask a leftist whose Indigenous territory we are on and nobody really knew. It was never common at rallies to have a speaker from the local Indigenous nation come to the rally, to open it up and to welcome people. Stuff like that never happened.

It was only in the 90's that it began to happen in Oka, which was a confrontation between a Mohawk Nation and the Canadian state police and military. It was a major standoff that lasted 77 days in the summer of 1990. After that, there was a big change: in a lot of cities, especially in Vancouver, there was a big push to understand whose traditional territory you're on and to acknowledge that at rallies and events. It then became very common that you have someone from one of the local Indigenous nations open up the event.

Anticolonial consciousness has increased a lot since the 1990s. The solidarity with Unist'ot'en is the result of all that anti-colonial resistance work. Especially in Canada, I think anarchists are a lot more on top of anti-colonial resistance solidarity, because of the importance of Indigenous peoples in Canada and the impact that they have on the society today. I think it's different in the U.S., because native people are a much smaller percentage of the population. You also have much larger Black and Latino populations, so a lot of efforts of the U.S. Euro left settlers are often focused more on Black and Latino community organizing. They're a much more significant political force in the United States than Indigenous peoples are. In Canada, there's a different kind of context here, but anti-colonial consciousness has grown a lot more since the early 1990s.

PATRICK FARNSWORTH: I recently interviewed Peter Gelderloos, who is an anarchist activist and author of *How Nonviolence Protects the State*. I really enjoyed hearing his perspective on how nonviolence is often very limiting. I know that you can only really speak for yourself in what you've done as an individual within these movements, but what is your general opinion on what's most effective? What has been most effective for these Indigenous resistance movements?

GORD HILL: People think it's a cliché or a nefarious thing, but: diversity of tactics. That's what works. Oka was an armed stand-off. You had Mohawk warriors armed with AK-47s, and they militarily confronted the police and Canadian forces. There were also, across the country Indigenous people mobilizing in solidarity with the Mohawks at Oka, and they used a large diversity of tactics. You had rallies, occupation of government offices, highways and railways being blockaded, as well as sabotage of infrastructure.

That's what occurred, and that is a diversity of tactics. That's what prevented the Canadian state from using military force to remove all the warriors. The military were very restrained in their actions because of the solidarity coming from Indigenous people across the country. The Canadian state feared an Indigenous insurgency could arise if they used deadly force on the Mohawks. And you have to remember: at the beginning of that standoff there was a police officer who was shot and killed. Indigenous people didn't condemn the Mohawks for defending their land militarily, though it had resulted in the death of a police officer. People rose up in solidarity with the Mohawks.

Nonviolence and pacifism are fairly recent ideologies. It's a belief that your nonviolent tactics and strategy are morally superior, presenting nonviolence as a better way. They claim it's morally superior, and then claim that it's more effective. But in reality, that's not true at all. Most of the nonviolent victories that have been claimed by pacifists actually involve a diversity of tactics.

Going back to the Mohawks at Oka: they used violence. They were armed. They dug trenches and barricades. They shot at police. And there was also a diversity of tactics that limited the ability of the Canadian state to actually repress that. But in the end, the Mohawks won.

The fight started over the township of Oka, which wanted to expand a golf course and build a condominium project over the pines, which was a Mohawk burial ground and a very important, sacred area to them. That's why they rose up. They were trying to stop this development from happening, and they won: the golf course was never expanded, and the condominium project was never built. There's many, many examples of Indigenous peoples using a diversity of tactics—including violent means—and achieving a victory.

That runs contrary to the pacifists' claims. Pacifism is a recent thing, aside from some religious cults that preach nonviolence. Pacifism only really arose in the 20th century, in the 1920's and 1930's with Gandhi. And Gandhi was promoted by the British, who were the imperial power that ruled India. So, Gandhi was almost an agent of British colonialism, carrying out counter-revolutionary activities in India to suppress an actual revolutionary movement.

India saw massive campaigns of rioting, guerrilla warfare, and assassinations of British officials. At the end of the day, the pacifists will still say that India got its independence through nonviolence, because they're stuck in their belief that pacifism is morally superior to all other forms of action.

So that's my take on pacifism and nonviolence. It's promoted primarily by middle class entities. Pacifism was never a part of our Indigenous cultures. When we were faced with a physical threat, we respond with physical force to defend ourselves. That's common sense. That's what any living being does. It's just natural.

It's only with pacifism that you have this belief that we have to respond only using nonviolence. They'll present it as politically practical. Of course, there are times when you just can't use a militant force to carry out something. That's why you have a diversity of tactics.

PATRICK FARNSWORTH: Something that Peter talked about in his book was the idea that if you engage in violence, you're going to only perpetuate violence into the future. By "dirtying your hands" with violent tactics, you're going to perpetuate violence. It's sort of a strange logic, because it's really obvious that that's not the case.

GORD HILL: It's the moral judgment that a lot of pacifists carry with them. The actual roots of pacifism are in a religious cult. Gandhi's whole project was basically a religious cult, and Christians come to their pacifism through religion. That's why most pacifists are so moralizing and judgmental of other people, and think they are superior. That belief has a religious basis.

I'd like to acknowledge a few things. First, less centralized, more horizontally structured forms of organizing are crucial in building resilient communities and networks. Second, accepting a diversity of tactics and dispensing with the false dichotomy between so-called violent and nonviolent action is necessary in directly confronting the systems of power. Third, we must learn from each other; each form of resistance to this murderous, life destroying civilization must build on each other.

Solidarity is key in building and sustaining any form of effective resistance to the vast Empire that will use any and all its expansive resources to crush any form of resistance to further its exploitation of the planet. Regardless of how obvious it may be that we have crossed numerous thresholds, the logic of this system will not yield, under any circumstances. Preparation is key in this fight, and all the tools we need to resist are right here in front of us. It's just a matter of knowing how to use them.

XABAT

Since the outbreak of the civil war in Syria in 2011, much of the focus (at least in the media) has been on a few key dynamics of this conflict. The first is the brutal nature of the Syrian government's response to the protest and resistance against the Bashar al-Assad regime. The second is the geopolitical implications of various world powers and regional players attempting to influence the outcome of this war in their favor. The third is the extraordinary rise and fall of oppressive fundamentalist Islamic militias and groups, like the so-called Islamic State (also known as ISIS or Daesh) and Al-Qaeda affiliated cells (the "moderate rebels") in their attempts to carve out their own territories in the region more broadly.

Out of this horrendous conflict, there has been at least one major positive development that has emerged—the Rojavan Revolution. Deeply rooted in decades of Kurdish militant resistance against state oppression and capitalism, the civil war and the fracturing of Syria has allowed space for a deeply feminist, multiethnic, and ecological society to emerge in northeastern Syria. This society has worked very diligently to build, maintain, and expand egalitarian social structures that collectivize decision-making on multiple levels, with a particular emphasis on upending patriarchal systems and transitioning away from capitalist modes of production in all aspects of economic and social life.

In March 2019, I interviewed Xabat from the Internationalist Commune of Rojava, situated in the Kurdish held territories of Northeast Syria. Speaking from the heart of this revolutionary movement, Xabat provides a glimpse into what a contemporary, militant, and deeply revolutionary society looks like (with all of its inherent flaws and contradictions) in the midst of civil war and ecological crisis. Much of this discussion is framed within the subjects addressed in the book *Make Rojava Green Again*, released by the Internationalist Commune of Rojava, which examines the environmental perspective and ecological practices inherent in the egalitarian and anticapitalist revolution in the region.

PATRICK FARNSWORTH: When decisions need to be made about how to use the soil, how to reforest, how to bring about biological diversity in the region, how are those decisions made? How does the ecological dimension of this revolution fit within this egalitarian social model?

XABAT: First of all, the three principles of this revolution are an ecological society, women's liberation, and radical democracy. So already in the pillars is this aspect of ecology, overcoming the annihilation of nature.

But of course, we have to look at how it is in practice and daily life. It's a difficult process, it's a revolution. So that requires changing also the mentality of the people, to change the way people are thinking, and how they address nature.

One of the big impacts of capitalist policy is that society got disconnected from the land. You see this in desertification and the loss of knowledge about producing our own food. Over the years all this was lost with the centralized nation state, and it got lost in society also. This is something we have to claim back. We have to recover it again here. What were traditional ways of production, of agricultural production here in this region? What kind of traditional crops were growing here? What trees were growing here? All these things have to be recovered. The knowledge is still in the society. We just have to dig it out again.

In the radical democratic way of making decisions, the smallest entity is the commune. It can be one street in a city, and in the village it can be one neighborhood. So really on this level of society, people are discussing the problems they are facing in daily life, trying to find solutions. And if they cannot reach solutions, because maybe the problems or the questions are related to the city, to the municipality, or to maybe even the whole canton, then questions and discussions are passed on to the next level where people from the whole neighborhood or from the whole city take on the question. So this is the context of the ecological work.

THIS REVOLUTION IS REALLY SOMETHING FUNDAMENTAL: NOT TO IMPLEMENT LAWS OR USING FORCE TO CHANGE THE WAY OF PRODUCTION OR THE STRUCTURE OF SOCIETY. INSTEAD, IT'S BY LONG-TERM DISCUSSION.

We are facing the problem of the awareness of people around these issues and also the economic aspects. To diversify agriculture also means to find how the agricultural system works right now, and how need influences this. There's a need for wheat in the region, and it is really easy for farmers to produce wheat. It's also riskier for them to change the system right now if they just see their personal interests and their personal situation. The work is also to change the awareness of the people.

This revolution is really something fundamental: not to implement laws or using force to change the way of production or the structure of society. Instead, it's by long-term discussion. In some villages, you have to discuss for years before implementing a new ecological and cooperative way of production. But it works: we can see many examples where after this long process there are cooperatives for working on the land and domesticating animals. This is the way the system here is functioning: it takes a long time, and it needs a lot of patience for all of us.

This is the general understanding, not only in the context of the agricultural aspects but also for example in the case of water. The committee that is responsible for this area counted all the wells to get an overview and implemented watches to see how much water's actually used to understand what is needed and how we can reduce the use of water. Now, you are not allowed to open new wells without

getting registered, and so there's actually an overview about it in some areas.

Also on the control of livestock, there is an implementation of natural reserves, and sometimes there are conflicts from one day to another. Farmers cannot use their monocultural way of production in this area anymore, because it's forbidden to use some kinds of fertilizer, just as it's forbidden to use pesticides. So we have to find solutions for a more ecological way of production. But in all these cases, we tried to do this process together with the most basic levels of society, the people who are first affected by the questions and they actually lead the process of finding new ways and solutions for the problem.

Rojava stands as a beacon. Under decades and centuries of capitalist and colonialist exploitation are the sprouting seeds of something so liberatory as to be hardly contained. Rojava is a particular manifestation of this spirit, and to me, represents a convergence of many of the broad struggles discussed in this section up to this point. There are many Rojavas, or at least there could be. One can be destroyed, but many Rojavas the world over would stand a far better chance against the world-eating machine of Empire.

This is idealistic, isn't it? In a book that deals with such prospects as near-term human extinction, presenting any semblance of idealism in our time of compounding crises and ecological apocalypse may seem like a disingenuous move. For me, though, it is quite simple: this is the work required of us right now. We build revolutionary societies and movements not because we have any delusions about it "saving" us or the world from destruction (and if that ever was the case, that time is past). We do it, because what else are we going to do? Stand idly by as the world burns?

The world is already burning, and the flames will continue to burn hotter and more destructively, regardless of what we do right now. In spite of this, how we direct our efforts is of utmost importance right now. Beauty is not the ideal, it is the lived experience.

RAMON ELANI

Let's try to frame this a bit differently now, because really, this is about our home. And to clarify what I mean, I want to present my interview with heathen writer and poet Ramon Elani.

When I try to get at the vast moral dilemmas we are forced to address in the face of these converging crises, confronting them through collective action seems apparent and obvious. While that may be true, another truth emerges from the prose of Ramon. He doesn't shy away from certain truths implicit in modernity and the reality it has forged in its attempt to subject the Earth and its living systems to the logic of "progress" and endless growth.

In Ramon's work, the home and our relationship with the land becomes the center of our resistance to the life-destroying forces despoiling the planet.

PATRICK FARNSWORTH: I've been enamored by your writing over the past several months. It took me a little while to catch on to your writing, but you're putting out this extremely valuable work. Your writing, in the prose that you use, is incredibly captivating to me.

I'd like to quote a section of one of your essays and ask if you'd discuss it more.

> Make the home the center of your life. For so many, home is lost in memory and dream. Nostalgia, a devastating longing, a force beyond our ken. What is it about the idea of home that is at once so comforting and so uncanny? It is precisely that link with those subterranean currents within us. The notion of home so often retreating from us brings us back to long forgotten memories. Not of our own childhoods necessarily, although of course the dollhouse world of the child is the model for the great world beyond, but the life of meaning and connection severed by industrial society.
>
> For where else is the power of the pre-modern world felt more strongly than in the home and in the idea of the home? We cannot return to our home any more than we can return to the wholeness that was taken from us. But we can reclaim something of our inheritance. We may light the fire in the hearth, call the gods and spirits to us, and make a new home for ourselves.

RAMON ELANI: An underlying theme in all of my thinking and writing is the meaning of modernity or the modern. Modernity is one of these kind of famously contested terms, but for me, modernity is a particular conception of time. It's a temporal concept, it's a concept about space and movements, and so it is a fugitive thing. Modernity is something that *runs from*.

It's also an experience of homelessness. When the tendency that modernity represents became most full in its manifestation—sometime in the sixteenth or seventeenth century—it was driving people from their homes, both in the sense of displacements and in traveling and adventuring.

The nomadic existence of modernity is very different from the nomadic experience of the premodern: you certainly have nomadism before the sixteenth century. But traditional nomadic peoples are always going in a circuit, from point A to point B and then going back to point A. This circuit or cycle represents all kinds of things, and it's not hostile to the idea of home. But modernity is homelessness in the sense that it is always driving us out. It drives us away from the land, drives us away from being connected or tied to the land.

In that essay, I was particularly thinking about some of the writings of Wendell Berry, an enormous influence for me. For Wendell Berry, agrarianism properly understood is not industrial agrarianism, but a philosophy of the home. This is something that we see in Rudolf Steiner, too: a farm properly understood is a home, an ecosystem, and an economy. And of course, the word for economy just means the business of the home. For these thinkers, the idea of the home implies a relationship to the land.

Also, one of the overwhelmingly prevalent features of pagan belief systems is the hearth, the hearth spirit, the hearth goddess, and ancestor worship. There is a spiritual dimension to the home; the home is not merely the place where all of our possessions reside. It's not merely the place where we work the land, but it's also the place where our relation to the spiritual is most alive. So we see hearth goddesses and household spirits, and the idea of household spirits is one that's always really resonated with me.

Carl Jung at some point left the city and built a stone tower by the side of the lake and lived there for some time in a very primitive way, which is to say, with very rudimentary technology. He talks about cast iron pots and pans, and for Jung these were pleasing to the gods of the house and home, while modern kitchen gadgets and so forth are alienating to the gods of the home. So that's something that's always been very meaningful to me as well.

ONE OF THE OVERWHELMINGLY PREVALENT FEATURES OF PAGAN BELIEF SYSTEMS IS THE HEARTH, THE HEARTH SPIRIT, THE HEARTH GODDESS, AND ANCESTOR WORSHIP. THERE IS A SPIRITUAL DIMENSION TO THE HOME.

PATRICK FARNSWORTH: What you're speaking to, I think, addresses something much deeper than what most people in our modern culture or society really understand the home to be, and also points to an unease with the idea of home.

RAMON ELANI: Yeah, absolutely, an unease with both our actual homes and our concept of home. I spent my whole childhood living in apartments that I didn't stay in for more than a year at a time. I was moving constantly as a child and as an adult. I think for most in our generation, home has become just a practical and basically meaningless concept. It's the place that you stop at in between work and whatever else you're doing.

Whatever you're doing that feels meaningful, universally it is not happening in the home. For most of us and definitely for myself until recently, everything that was important about my life was something I did somewhere else. This is something I think about a lot in terms of the relationship between the traditional and the modern, that really with very, very few exceptions we are all homeless now, really all of us, even those of us who live basically in the same place that the ancestors did.

One has to make one's own home at the end of the day. I think that we have to find our home as individuals and as a species if our species has any future. I think home is something that needs to be created, that's the point I'm trying to make.

PATRICK FARNSWORTH: If you don't mind, I want to read one more part from that essay.

> The human world is in ruins. It will not get better. The sooner we can withdraw from it, the better. Timothy Leary was right when he urged young people to "drop out" in 1966. His message is all the more profoundly true today. Life in urban, industrial society has no future. The modern world has failed on all levels. Capitalism and industrialism cannot be reformed. The gods have fled. Whether or not we can become completely independent of industrial society is irrelevant. The fact that it is difficult and perhaps impossible to utterly separate should not be used as an argument against withdrawal. Connection to the gods and the land is ultimately more important than material self-sufficiency. To whatever extent you are compelled and able, withdraw from society and make the home the center of your life.

A lot of people who are writing about the hollowness of industrial modern life suggest we need to organize to resist it or to undermine it. In this particular essay, you're saying: turn your back on this society, because it's not going to serve you, it's not going to be fixed, it's not going to be reformed, and it's not going to get any better. It took me a little bit aback, because others suggest we have to do this in a collective sense but you're pointing to more of an individual choice.

RAMON ELANI: I've come to where I am as a person through a meandering path. Most of my adult life, I would have been horrified to hear myself say this, but that's no longer the case. It's taken me a long time to come to the conclusion that reform is impossible. This society cannot and should not be saved, and our efforts would be better placed doing other things.

The suggestion that I'm putting forth is to make a home for yourself. I recognize that making a home for oneself can mean a lot of different things, so I'm not trying to be prescriptive. But absolutely we've got better things to do with our time than to try to save the society.

It's such a truism that the Earth is our home. We hear that all the time, and we are comfortable with that formulation but ironically, we very rarely meet somebody whose idea of home is meaningful to them. That clearly is a problem. We've spent 30 years, maybe even less, seriously wrestling with the fact that we are destroying the earth, and yet I don't think very many people have a clear sense of why it should be saved or even what saving it means. People are very good at responding to things that are immediately threatening them, and if you're under attack then resistance is of course understandable. I'm not suggesting that those who are attacked should not resist.

But it's always been very funny to me we are saying that the Earth is our home, so we should protect it. But if our homes are just meaningless places where we dump our possessions, then why is it important to save that? That doesn't sound very valuable. I think the problem is a spiritual one, and it's really about the home and our connection to the earth. And that's something that the individual has to sort out for themselves.

PATRICK FARNSWORTH: What do you mean when you say making space for the gods, making a home for them to dwell?

RAMON ELANI: The philosopher David Abram, who's written two amazing books on animism and phenomenology, has this great passage in one of his books where he says that we assume that trees do not speak in modern society. He suggests that we think that trees don't talk because we don't talk to them. Why would anybody talk to somebody who doesn't talk to them? If we're not talking to the trees, then what motivation do they have to talk to us?

I think this a great way of conceptualizing animism and spirituality in general. I was raised as a hardcore atheist. That was the context in which I was raised, and it took me a long time to realize that not only do I not feel that way, but actually I don't think there is such thing as atheism at all. I don't believe in atheists. I think that everybody that consider themselves an atheist is actually something else.

Freud had an exchange with a French writer named Romain Rolland, who is important for a number of reasons, but mostly because he brought mostly the Vedic philosophy to Europe in the early twentieth century. They had an exchange around Freud's book, *The Future Is an Illusion*, which was his big takedown of religion. Freud knew that Rolland was very spiritual, and so he asked Rolland to do a review of it. Rolland wrote back that he thought it was really good, but he felt that Freud overlooked the initial motivation for religion, which is something that I think a lot of people don't consider. What Rolland suggests is that there is a sensation he believes is universal, and he likens this sensation to floating on the surface of the ocean. He calls it the "oceanic feeling." It's simply this sensation of being supported by a vastness below and around you, of which you are a part but also discrete from.

So you're kind of just floating, kind of bobbing along on the surface of this vast sea. Rolland says that he's experienced this many times, and that he thinks that when people are talking about religion or spirituality, they're trying to put a name to this sensation.

I would say is that I came to my understanding of the gods through the recognition of that sensation. Unsurprisingly, that sensation occurs far away from human things, far away from human beings, human places, far away from human stuff. The only way I can describe it is that I heard them speak to me and I spoke back to them. I speak to them constantly. I constantly have an ongoing dialog with the gods, which I name in various ways.

This is from Jung, but one of modernity's initial assumptions is everything that you've experienced spiritually is superstitious nonsense. This is one of the primary defining characteristics of modernity: dismissing all of that stuff as rubbish. This is also why modernity has been very accommodating of Protestantism, because Protestantism out of all religious systems is civic-minded, practical minded, and economically minded. Capitalism probably couldn't have occurred without Protestantism. and so on and so forth.

When I talk about speaking to the gods, naming them, and acknowledging them, it's also a way for me to acknowledge forces and powers that are beyond humanity. I think that the key thing is that the relationship with the land is a spiritual act, because it acknowledges agency. And again, you've got to have somebody to talk to you if you suppose that there's going to be a relationship.

Many people would argue that retreating from society is not going to help anybody. I'm of two minds when it comes to this. Ramon's perspective can help us reframe our approach to activism in a very helpful way. Focusing on the Home, in the profound sense Ramon points to, can ground us and further clarify why we fight.

This question must be asked: if we don't even treat our homes as sacred, how then can we even approach treating the earth as sacred? Our resistance must come this place. Along with engaging in the broad struggles and forms of resistance addressed in this section, with all the very complex questions and perspectives that come up, for me, it's always about the Home. Not just an abstract concept, but it is the direct, lived experience of the sacred. The Home is what we are fighting to attain and protect. Modernity has stripped this from us, and this is why the destruction of the planet proceeds uninterrupted.

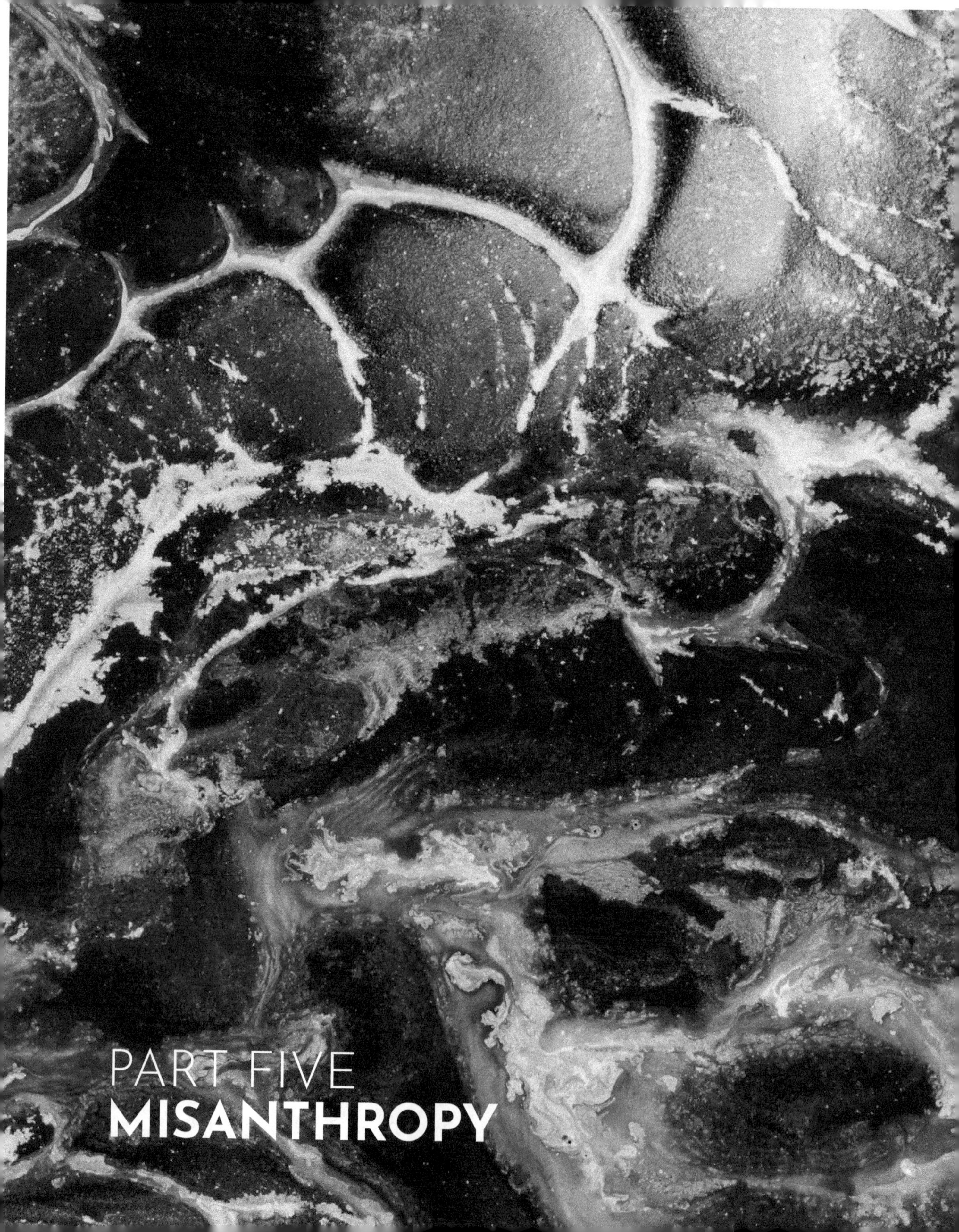

PART FIVE
MISANTHROPY

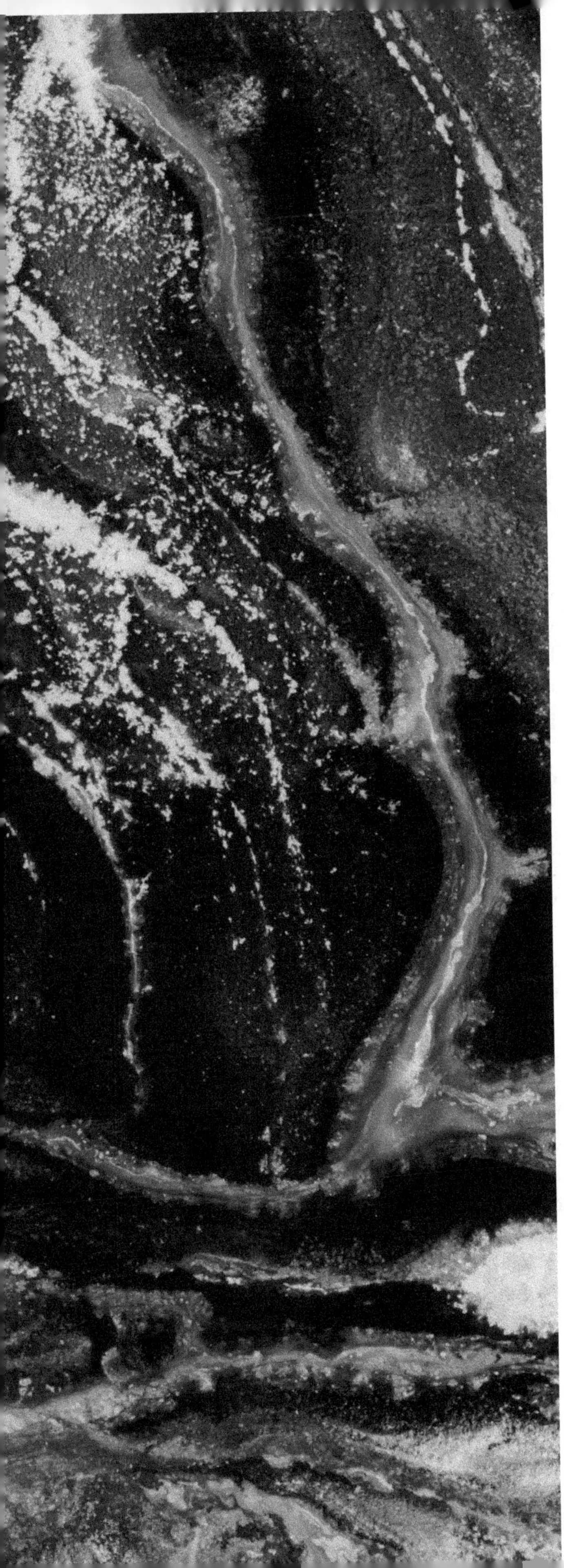

RESISTANCE TO THIS SYSTEM is crucial, and knowing *why* we resist is just as important—if not more so—than the kinds of tactics and organizational forms we use. Certainly, we need to resist, organize, and build something better in the time we have. But the simple question "why" we are resisting can orient us toward our ultimate goals and help us understand what our principles are in this broad struggle.

The nature of the crisis is massive, and the sense of overwhelm we experience when we

grasp its scope can be debilitating. This feeling is common, one to which I deeply relate. Years of soul-crushing depression and despondency regarding our predicament has helped me recognize the full spectrum of reactions we can experience to this perilous moment.

I cannot adequately explain how I managed to move through my own Dark Night of the Soul, because it certainly wasn't any particular thing. I can't clearly define it, but something managed to pull me through it. Suicide was certainly something I considered. It's somewhat difficult to admit, but I needed to engage with nihilism and misanthropy to recognize what a dead end those paths really were. I learned that hatred for humans, believing them to be a no more than a virus or scourge on this planet, is really just another form of self-hatred. It serves nothing, except to feed "an incandescent rage that poses as a conscience," as Stephen Jenkinson would say.

In this section, I will attempt to engage with a wide range of responses to the predicament we are in, loosely defined as "false responses" and "false solutions." These responses are unproductive at their most benign, and extremely dangerous at their worst.

To reiterate, this isn't about finding hope in the sense of "winning," but rather orienting ourselves by our sacred responsibilities. What these sacred responses may look like will be elaborated more fully in the last section, but it's imperative we acknowledge how some of our potential responses can be truly beautiful, and so many others not.

We must prepare for the worst. There are historical, cultural, and cognitive reasons for these crises as well as these responses. We can hold space for the sacred while fully acknowledging the horrors of the human condition and the vile responses that are emerging in this time of deepening despair and wild reaction.

MICHAEL SLIWA AND ROB SEIMETZ

I'd like to begin this exploration with a segment of my discussion with Michael Sliwa and Rob Seimetz. Rob is a writer and former host of the program *Moving Forward*, and Michael is a retired public school teacher, former cohost of the *Nature Bats Last* program, and the author of *Chasing a Different Carrot: A Manifesto or the Predicament of Privilege*.

In our discussion, the three of us examine the overwhelm many are experiencing trying to navigate through this time of overlapping social, economic, ecological, and political crises. We focus particularly on what can be loosely defined as "doomer" communities. These have sprung up from the growing awareness of these intersecting crises, in particular the climate crisis. Each of us are quite familiar with the ideas surrounding "near-term human extinction" and the data that backs up this view, some of which I presented at the start of this book. None of us deny this to be the case, but we are somewhat critical of the more misanthropic strains that tend to run through the community, as this view severely limits us in our response to the crisis.

PATRICK FARNSWORTH: When I think about what doom and collapse actually look like: I don't know. I don't know if anybody really wants to go there. But regarding those who take any glee or any real joy in spreading that feeling—there are two ways to do it. There's the glee, and there's the self-hatred, the human-hating. The human-hating aspect of it is more of an arrogance, like, "I know better than you, I'm an elitist."

But then there's another way: "this is the physical reality that we're all in. I want to break the news to you, but I'm not necessarily happy to do it. We need to figure this out together and I'm going to try my best to convey that information to you in a blunt but also in a rather sensitive way as well. And I'm not perfect at it, because there's no guide to this."

MIKE SLIWA: I think both of you are very good at doing that. You're much better at doing that than I am. I came out into the doomer world with a master's of hitting people with a sledgehammer and delivering the news in that forum. I had a student visit me recently, one I had as a freshman. He's, like, 25 now. But he was a kid that was very deep and very curious. And talking to him was like talking to an adult. He could have these conversations, I thought at the time (this is around 2008). I was just "getting my doom on," so to speak. I was reading *Endgame* and all that stuff.

We would have these conversations. And then he told me, later, after high school: "for two years after that, I just hated you, because you had delivered this devastating news to me. I was just trying to get through high school...."

By his senior year, somehow amazingly, he was in a place where he could actually reflect on it. He did a video of me for his journalism class, and we talked about all that. He asked, "did you ever have any regrets about teaching?"

I told him, yes: that part. It was one of those things I was processing through myself. I told him, you were very inquisitive and you had great questions. You had great answers. And I probably should have checked myself about some of those things, because we were talking about heavy stuff with a lot of students, about racism and patriarchy.

We did some amazing stuff at this school. But in reflecting back on those things, I've said some things over the years on podcasts and to students that I am much more reflective and careful about now. I let people approach me and show me where they are now before I start delivering shit like that.

PATRICK FARNSWORTH: Yeah. I think we have to give ourselves a little bit of leeway. I think we can be really hard on ourselves, especially in retrospect. Mike, I don't think you knew what the best possible way to convey this information to people was, or who you were at that time, and what you were experiencing, and how you were processing that information. For a while, that's how I wanted to approach it. I wanted to be like "we're fucked! Everybody, we're fucked!"

But nobody listens to you when you do that. Nobody takes you seriously, even when you present scientific data. This is a fact about human beings—in general, presenting data and presenting scientific information doesn't convince people...

MIKE SLIWA: It's worthless.

PATRICK FARNSWORTH: ...of anything necessarily. If they're predisposed to believing that stuff, then maybe. But if that many people don't, that's not how people change their mind at all. And I think...

MIKE SLIWA: It's got to be a narrative.

PATRICK FARNSWORTH: Yeah! It's absolutely about a narrative isn't it?

ROB SEIMETZ: But I do think, with more of those studies and more of the science out, there is a tide turning among people that believe in climate change and believe it's a threat. I think it's up to 70 percent now, which is a lot higher than what it was. We have a very unique opportunity. What are we going to do with this rising tide of awareness?

I think the right's going to try to seize on it eventually, too. So we've got to be really careful. One of the things I think about is that everyone's at different spots, but I'm at the point now where I get the science—but now what? After the science, what do I do now? And also, what are my passions? I am interested in other people who have decided this is where they're at. What are they doing about this, and what are they passionate about? Because I feel like that's what we're going to learn things from.

For instance—he was a guest on your show, Patrick, and on mine—Dean Spillane Walker. He's doing this Deep Adaptation course, he's trying to get into language and how we talk to one another with this going on. That's the type of work I'm interested in. That's the type of work I want to gravitate to.

PATRICK FARNSWORTH: Yeah, I think people are coming into this information and then the question that they're going to have after they start that grieving process is, "okay, so now what?" Are we doomed to nihilistic despair, is that really how we want to go? No. That's why I want to push back against elements within "the doomer community" like that. Because, really, is this how you want it to be? Is this what you really think this whole thing has been about? I understand it's really disheartening. Obviously, I am disheartened all the time. But I'm not going to sit here and "doom" on everybody, because they're human beings.

A person that I've talked to—I just released an episode today with him, my second interview with him—Joe Brewer: he is somebody who is working on building regenerative hubs. He moved to Costa Rica with his family this last year, and he's working on building regenerative systems and asking, how do you design cultures? How do we make cultures that incorporate the ecosystems and bioregions that people are in? How do we build those?

He's incredibly intelligent. He's multidisciplinary. He's really amazing. His intelligence is astounding, and something he said that really stuck with me I'm thinking about it now is, "what can we do that is worthy of our hope?" That stuck with me, because there's so much of this talk of "hopium." Like, don't give people any hope because we're all doomed, and doing that will just make people believe that there's a solution. Joe is working on "solutions," but with the full awareness that it's probably not going to work, that we're on a path towards our own self-destruction, that the planet is going to be irreversibly changed, and that we're not going to be able to regenerate it. That doesn't mean that he's not doing something.

People want to do things based on an outcome: "I'm going to clock into my job because I know I'm gonna get paid for it. I'm gonna do this activist thing because if I go to this rally, it's gonna make governments change their policies." We shouldn't be doing it for that reason. We should do we doing it because we love the earth. No species on this planet has deserved what it has been given as a result of human industrialization. None of them have. And we don't deserve it either.

ROB SEIMETZ: I think when you hear about these scientific studies and you read stories about the rising far right politics and fascism across the globe, and the fact that we have a white supremacist for our president—if somebody doesn't believe that, they're hiding under a rock. The story last week was that he's okay with people shooting immigrants that try to cross the border. All these things to me are reminders and nudges on my shoulder that the planet is asking you to do something right now. And that's a different question that each of us need to answer. And I'm trying to figure out how to answer that question for myself. Everyone's going to go about it differently.

MIKE SLIWA: Absolutely.

I think part of it is that we're all so conceptual, and there's not a lot of "actual" going on and it sounds like Joe is in the "actual." He's actually doing something about it. Whether or not it's going to come to anything, he's doing it because of a sense that "I have to do something." And I think we tend to overthink it, at least in the broader general population.

But in the "doomer" community, we like to think of the conceptual, about how this is all going to go down. What am I going to do if this happens? If the hordes start coming after us for our stuff? Instead of actually asking, "what can I do today to make my day better, to make the

YOU CAN'T CONTROL WHAT'S GOING TO HAPPEN, YOU'RE NOT GOING TO BRING DOWN THIS INDUSTRIAL MODEL ON YOUR OWN. BUT YOU CAN MAKE CHOICES IN YOUR DAILY EXISTENCE THAT WILL KEEP YOU SANE AND ALLOW YOU TO BE A LITTLE GENTLER WITH OTHER PEOPLE.

people I interact a little better, and to have less impact on the planet?" It's not going to save the planet, just like my living in a tent in the middle of nowhere is not going to save anything.

It changes your lens though, changes how you interact with not only your own species but with other species. That's all you can really do. You can't control what's going to happen, you're not going to bring down this industrial model on your own. But you can make choices in your daily existence that will keep you sane and allow you to be a little gentler with other people. It's taken me awhile, and I've still got a lot of rough edges. But being out here has reflected back to me the asshole that I've been through a lot of my life.

When you start to realize through the years that you only get one of these lives, you might as well be nice to people, and nice to the other species. And try to go through this life without creating too much of a wake in your path.

ROB SEIMETZ: When you and I talked to Dahr Jamail, I could just feel his presence, the warmth, the genuineness. I think that's what is important, that we have conversations with people, that you get to meet people in person that maybe have similar views as you. I always have walked away from interactions with them and say, "I need some of that for me." And I have a lot of work to do.

I want that. The thing that gets me, when we get back to the Puritanism of this, is that because he's changed his thinking, people are crapping on him for that. This is just astounding to me. Talk about a steward of the earth.

PATRICK FARNSWORTH: Yeah, I think the Purity Test is definitely present. I want to say something nice first, before I shit on the doomer community again. I actually see very often people who come online, who come into these groups and are like, "I don't know what to do. I'm overwhelmed." And there's a lot of overwhelmingly supportive people out there, so I don't want to just shit on them.

MIKE SLIWA: Oh yeah. Absolutely.

PATRICK FARNSWORTH: I'm a part of this community to some degree, and I see great people all the time. What I see also is a pressure on someone like Dahr, who has a best-selling book. He messaged me one time, I think it was for the event we saw him at Powell's Books, and I think afterwards he said, "was that doomy enough for you guys?"

MIKE SLIWA: I think I'm partially responsible for that. *Nature Bats Last*, in the beginning as far as podcasts go, set the level of what was acceptable and what wasn't in the doomer community. All the extinction groups reinforced a lot of that, and the whole "hopium" thing? That was us. Those are things that I am partially responsible for putting out there, people measuring themselves to see if they're dark or doomy enough.

You don't realize sometimes the power that you have when you have a podcast. We have several thousand listeners, and then other people listen to them years later. And then, people start following your writing. Both of you know this. It's a responsibility.

I am in awe of the people that reach out in their vulnerability, and the people that respond to them in kindness and with empathy and take the time to do it, because I'm generally not the person doing that. I'm not giving a lot of advice or hope or empathy to people. I don't want to overlook the folks who do that. A lot of times you look at the negative things in life that stick out, but there are a lot of really good people doing amazing things in those groups.

A thread I want to follow in that exchange with Michael and Rob is the range of responses we can anticipate as things continue to deteriorate. Already, people are finding each other in order to provide something like solace, validation, and support as we awaken to the dark knowledge of the climate crisis and the inevitability of widespread collapse. This trend, I think, is inherently good. We need these groups, and there are some truly incredible human beings that have carved out spaces for the support we need. Rob, Michael, and I acknowledge this in that exchange.

However, there is something I have become increasingly concerned about within some of these circles that I want to push back against: the human-hating component. Inevitably, the environmental crisis brings up very serious questions of how we have come to this point. This cannot be easily reduced to phrases like, "humans are a virus on the planet," as much as we may want to.

I'm deeply convinced this isn't the case at all. That logic is a dead end, even if we acknowledge the tragic quality of the trajectory we are on. Worse, this view is tied to too many deeply held, but ultimately flawed, assumptions about the human condition and what can be called "human nature."

JOHN H. RICHARDSON

To follow that aspect, consider my interview with journalist John H. Richardson. He addresses the more extreme paths certain individuals and groups have taken in resisting industrial culture and its destruction of the living world. John is a former "writer at large" for *Esquire* and authored the truly captivating article "Children of Ted: The Unlikely New Generation of Unabomber Acolytes," published December 2018 in *New York Magazine*.

In researching and writing this piece, John took a deep dive into the world of what he has labeled the acolytes (and apostates) of Theodore Kaczynski, more commonly known as the Unabomber. John documents his interactions with various individuals and groups that have been inspired (or adjacently inspired) by the anticivilizational writings and philosophy of Kaczynski, as well as his multiple acts of terrorism that occurred between 1978 and 1995 resulting in the deaths of three people, with 23 injured.

Kaczynski's bombing campaign only ended after he struck a deal with federal authorities, promising to "desist from terrorism" if a major publication, like the *New York Times* or the *Washington Post*, published his manifesto unaltered, which they did in 1995. The FBI ascertained his identity and whereabouts after his brother recognized his writing style and the themes presented in the manifesto. His brother directed their investigation toward Kaczynski's residence, a secluded log cabin in the backwoods of Montana. Kaczynski was arrested in April 1996 and is currently serving a life-sentence in a maximum security prison in Colorado.

John's article, *Children of Ted*, centers on the story of primitivist writer and philosopher John Jacobi, who become radicalized (in part) through his exposure to Kaczynski's manifesto and his subsequent correspondences with him. Richardson examines a phenomenon he has described as "Kaczynski moments." These moments come about when an individual has read Kaczynski's manifesto *Industrial Society and Its Future* and then comes to recognize its arguments cut to a core truth about our culture: the impact industrialization has had on human societies, human psychological health, and the ecological destruction that ensues.

He's not wrong, at least in this regard. What is disturbing, however, is that this text was written by a man who condoned, promoted, and participated in acts of terrorism against individuals whose professions he perceived as promoting technological advancement and environmental destruction.

In the article, Richardson writes: "The Kaczynski moment dislocates. Suddenly, everyone seems to be living in a dream world. Why are they talking about binge TV and the latest political outrage when we're turning the goddamn atmosphere into a vast tanker of Zyklon B? Was he right? Were we all gelded and put in harnesses without even knowing it? Is this just a simulation of life, not life itself?" Kaczynski moments cut through almost every political tendency and ideology and yield wide and varied results, which Richardson explores lucidly in his article.

In writing the piece, John interviewed and spent time with some of the most prominent and infamous anticivilization activists and writers, and even spent time with a propagandist for the terror-friendly eco-extremist outfit Wild Reaction (formerly known as *Individualidades Tendiendo a lo Salvaje*, which translates to Individuals Tending Toward the Wild, or ITS for short) that is based primarily out of Mexico, which he elaborates on more in this segment of my interview with him.

PATRICK FARNSWORTH: I think what scares people is this feeling that this isn't actually under control. We are heading toward something inevitable if we continue down this path. And that's the feeling: waking up to that. And then the questions that come with that are terrifying for many people.

JOHN H. RICHARDSON: Yeah.

PATRICK FARNSWORTH: And I want to point to this, because we're talking about John Jacobi. You follow his story through this whole article. This is the thing that gets me. I hate the word problematic...

JOHN H. RICHARDSON: No, say it.

PATRICK FARNSWORTH: ...but I want to say there's a certain darkness that comes with this. You mentioned the journal. It's a collection of various essays called *Atassa*...

JOHN H. RICHARDSON: Yeah, okay ITS again.

PATRICK FARNSWORTH: Yeah. ITS is connected to this.

JOHN H. RICHARDSON: *Individuals with a Tendency to Savagery,* or *The Nature*, or *The Wild* or whatever you want to translate it as, yeah.

PATRICK FARNSWORTH: Yeah. And so Jacobi wrote at least one or two pieces for this journal.

JOHN H. RICHARDSON: Yes.

PATRICK FARNSWORTH: Later on in the article you mentioned you ended up meeting Abe Cabrera. And he also wrote a piece for that. So when you read it—and you say very, very clearly—it's a very terror-friendly publication.

JOHN H. RICHARDSON: [chuckle].

PATRICK FARNSWORTH: I mean, they're actually advocating to go out, to cause destruction, to go kill people. Human beings need to be destroyed.

JOHN H. RICHARDSON: Random violence.

PATRICK FARNSWORTH: Random violence, yeah, nihilistic: we're not trying to bring about an optimistic future. Kaczynski is something more of an optimist in a certain way. These people that are behind ITS, behind these groups, mainly come from outside the United States. These are people in Mexico, in South and Central America. And the people that are publishing *Atassa* are actively trying to stir the pot a bit and say: hey, all of your political theory is just bullshit, and it's not accomplishing anything. So we need to just lash out. And as they describe it, it's like they're channeling wild nature....

JOHN H. RICHARDSON: Yeah, they changed their name to Wild Reaction.

PATRICK FARNSWORTH: So they're bombing, they're shooting, they're killing people. They're attacking people in professions similar to what the Unabomber did. So in their acts, in their mind, they're the earth seeking revenge. They are the bloodthirsty, vengeful...

JOHN H. RICHARDSON: Animals.

PATRICK FARNSWORTH: ...animals, coyotes. And that's really disturbing to me because I don't actually think I can speak for the earth.

JOHN H. RICHARDSON: It should be disturbing [laughing].

PATRICK FARNSWORTH: Yeah. Yeah.

JOHN H. RICHARDSON: It is disturbing.

PATRICK FARNSWORTH: I've heard so much about *Atassa*, not just from your article. I know people from Deep Green Resistance. And I'm not associating them with this. I just wanna make that very clear. I know people that...

JOHN H. RICHARDSON: You know what?

PATRICK FARNSWORTH: From Deep Green Resistance, which you mentioned later on. And I'm not going to connect these two things together, but I'm trying to say there's all of these intersecting circles of people that have reacted against what *Atassa* is promoting and what ITS...

JOHN H. RICHARDSON: Oh, the anarchists hate it.

PATRICK FARNSWORTH: Exactly.

JOHN H. RICHARDSON: Anarchists hate the editor, Abe Cabrera. They took revenge on him.

PATRICK FARNSWORTH: Exactly. So, you spent time with some of these people. What was that like? I know you're doing it as a journalist, but what was so compelling is like reading this. I'm thinking: he's not just observing this. You're hanging out with these people. So what was that like? What drew you to that, that you were so compelled to go meet these people and contact them?

JOHN H. RICHARDSON: Yeah, I was trying to get in with the guys at Wild Reaction. And I thought Abe Cabrera was a conduit to that. I gradually began to realize that he was actually a chief propagandist in America for that. Or, not propagandist, but arguer for it. Abe is a really interesting character. He's a tortured soul. He's a former Catholic priest. He has an academic background. He has gone through all of them. And I found the most moving thing he said was that he's got children, and he doesn't tell them anything about anything he does. It's a double life. Just as Wild Reaction/ITS recommends.

I was never quite sure how far that double life went. But I think now he is a theoretician and not an actor. And the most moving thing he said was, "we're not destroying the world because we're greedy. It's not because of the oil companies being greedy. It's because we love our children and we want a better life for them. And so love is what is messing us up." I think there's a beautiful, tragic truth in that.

These people are unafraid of following something to its conclusions, and they are pursuing those conclusions. There's a validity to the ITS argument. It's unlikely that we're going to retool the economic system overnight—meaning in the next 15, 20 years—and stop doing what we're doing. Too many people's lives depend on it at this point. And it would lead to what people like ITS and Kaczynski unflinchingly see as a mass die-off. There's another guy in Europe, Pentti Linkola—he's calling outright for environmental fascism, where we control births and shut down everything but essential functions.

My interest in these guys is that all of them present a range of responses that are uncompromising and honest responses to what's happening. ITS broke with Kaczynski, they're apostates. That's why he hates them, aside from the fact that they put him in jeopardy when they mention him, because that's definitely a line the prison won't let him cross. They would shut off his ability to communicate, which is the thing he's most passionate about.

In the ITS sense, that's okay. They say, we're not going to win this revolution, we're not going to find a Lenin who can lead the people. As I told Ted in my post-journalist correspondence, "Lenin offered people real freedom and a more just world, but he [Kaczynski] is offering a hunter-gatherer lifestyle and mass die-offs. That doesn't ring the same populist message. It's a loser. As a propagandist, you suck." [laughing]

Even if that's where he wants to get to, he can't say it. Nobody's going to follow that. In the same way, ITS came to the conclusion that it's not going to happen, and they look to the Indigenous tribes that resisted the Spanish conquest and admire the tribe that fought the longest despite knowing they would lose.

This is also a tradition among the American Indians. They were split between the youth who wanted to use modern weaponry and the traditionalists who said "if we do that we're not Indians, we're not the Creeks or Navajos, etc." There's truth to both sides of that argument. It's tragic. We're all giving up our heritage.

That's where this intersects with the right-wing's concerns, in order to live in this world and function effectively and raise our children as Abe would say. I think there's a truth to what they're arguing: it's all horrible and it's not going to do any good, so why even bother? Why not just let it happen, and not be part of this violence and horror?

What's interesting is that these solutions are coming up. Paul Kingsnorth withdrew

from the climate change movement and gave me advice. I said, "I'm not a backwoods kind of guy. I'm not going to retreat to Northern Ireland and raise vegetables. I like the city I live in, New York City." And he said that there's plenty you can do there. He just believes that the global fight is lost, and that we can do the best we can with our communities.

There's a beauty to that argument. ITS says what it says, Kaczynski says what he says, Jacobi is still looking for an answer in his sweet earnest way. I like that, and I like John. I think more people are going to catch up with this, and I think as people see in undeniable ways that their farmlands are drying up and burning away, that their condos on the Miami beach front are going to lose value, and that their investments in various industries are going to start to evaporate—taking away their retirement savings and their kid's college funds—there's gonna be something of a wild reaction all across the world.

That reaction is going to go into some of these early forms. Maybe it'll find other forms. It blows my mind again that people don't see this. We will see a lot more of these things happening as the problem becomes more visible. We're already seeing this rise in right-wing fascism, the retreat behind borders—which is maybe not irrational when there's hundreds of millions of people on the move because of climate change.

So, if you're a scientist or a responsible government official, you should study it rather than deny what's happening. We put Eugene Debs in prison for protesting the First World War, a man who was the wonderful, decent, sweet, and idealistic head of the railroad union. We should have listened to him, because 20 years later we had the Great Depression and finally had a mass response that triggered some of the solutions he was looking for, though they were denied as soon as it was economically possible for a whole host of reasons. I don't want to just blame the capitalists, the media, and what did they call back in those days—tycoons or...

PATRICK FARNSWORTH: Barons.

JOHN H. RICHARDSON: Yeah, the barons.

You know, that's also a simplification, a simple radical feel-good kind of solution. I think all of those are denials of reality. If we're going to be honest about what's going to happen, we're going to start to look at some of these problems just as you would study disease. You study the outbreak of the disease, and the viral pathways through which it spreads. That's what epidemiologists do, and this is an epidemiological problem that's going to get worse.

I had pitched this story to a renowned activist publication involved in social justice causes. I admire it very much, I don't want to trash them because I love them. But their attitude was, "well, we don't think it's going to get that bad."

And my response was, "Have you read the science? Are you keeping up with stuff?" And the response that came back was, "Yeah, but we just can't go there."

Then I realized another one of these shocking things: the whole social justice project, that is so beautiful? If this stuff happens, it is going to become completely irrelevant and overwhelmed. And we can't look that in the face. I mean, if you're really coping with what has become the present problem, rather than a smaller problem or a historical problem, you're going to retool your whole magazine to this stuff.

It's like people who were worrying about new forms of agriculture when industrialism was taking the whole traditional farming way of life away, consolidating farms, which sent 70 percent of America out of their habitat. If you look what Wendell Berry wrote about in *The Unsettling of America*, this is a huge social wound that ties into all the stuff we're talking about that we're blind to. 70 percent of the country was working on farms and now it's 6 percent or something like that. I'm terrible with statistics but you know what I mean.

How can you not think about that when you're talking about the social problems of the city and right-wing resentment of centralization, and states rights and guns rights and all that? It's so much more complicated than people want to think. I just don't get it. I read Wendell Berry a while ago. So I've been putting the pieces together over probably the last 15 years. I also did a story in 2006 for no particularly conscious reason. I just thought it was an interesting story about an Earth First! activist who came out of prison and was just as relentless as he was before, which I found kind of admirable, Rod Coronado. I didn't agree with his solutions, wasn't compelled by them, didn't give a damn about animal rights, even thought it was a little foolish, although I understand it from other people's point of view.

So I was with him, and I met this kid who I really liked and I'm still in correspondence with who was just 19 at the time, and he kept talking about this guy Ted. And it took me a while to figure out who he was talking about it. And I said, "but he killed people." And he said, "yeah, but all of his hits weren't so bad."

I was like, "oh my God." I feared for him. I thought he's a really smart, really energetic, lively, and passionate interesting guy, and super friendly to me anyway. Maybe a little arrogant in the way 19 year-olds can be. And now he's got a family, he's working a job. But he still believes the same stuff. He just had to do what we all have to do.

He started mentioning Ted. And I just thought, wow, that's weird, and I didn't think about it again for years. But when I saw Jacobi's article, I remembered this kid, Zack, and realized I've heard this before. I started to see those guys' point of view, and I read Elizabeth Kolbert's *Sixth Mass Extinction*—a very respected journalist—and James Lovejoy, a great scientist: all of them were saying the same thing.

I had enough context by that time to think about it in a smarter, more informed way. The pieces started to come together, and I saw the patterns. That's how that first connection with a kid like Jacobi—though less troubled, from a more supportive family and all that, although I don't want to diss John's family, because I really don't know that story. Because I like Zack, I eventually came to really like John Jacobi, and even, a little bit, Ted Kaczynski.

I'm a sentimental journalist dude. I was unable to completely dismiss them. They were human beings struggling with issues that I began to see other people just deny. And I'm not hung up on them having the right solutions or the right interpretations. I've always written sympathetic pieces about right-wing figures too, because I try to see it through their

THEY WERE HUMAN BEINGS STRUGGLING WITH ISSUES THAT I BEGAN TO SEE OTHER PEOPLE JUST DENY. AND I'M NOT HUNG UP ON THEM HAVING THE RIGHT SOLUTIONS OR THE RIGHT INTERPRETATIONS.

eyes—they have their reasoning. I became pretty good friends, and continued to correspond for a while, with the head of Open Carry Texas. He had his reasons, they were coherent and he had a political point of view that I couldn't completely reject.

I feel I don't have to agree with their solutions or even their way of expressing it. I'm trying to look at why they're doing it. What's the reason? What's the thing that creates it?

I did a piece on Charlton Heston a lot of people thought was ridiculously sympathetic. He wrote a very nice letter to the editors of magazine. And then I saw another person whose work I admire, in Flint, Michigan...

PATRICK FARNSWORTH: Michael Moore?

JOHN H. RICHARDSON: Yeah, Michael Moore, whose work I like. I think he is another passionate, committed guy, and I agree with some of it and don't agree with other parts. I always basically liked what he was doing. He was ruthless to Charlton Heston. If you look at the way he came up, his mind frame—he's a decent guy, he's not a jerk or a monster. He's actually kind of a sweet man, or was. We got along, I spent a week with him. He's an actor, he sits there and talks about Shakespeare all the time. He grew up with those small town police sheriffs for a dad. There's a whole rationale to what he believes. I'm not interested in condemning, I'm interested in understanding or just seeing.

Anyway that's about me. But it's about all these people. You were right when you said there's really an overlap. And another thing: the left has taken over the climate change argument. Jacobi was making this case just a few days ago, very articulately in his way. Their hatred for the right blinds them to the fact that these people are going to be their allies someday, if they let

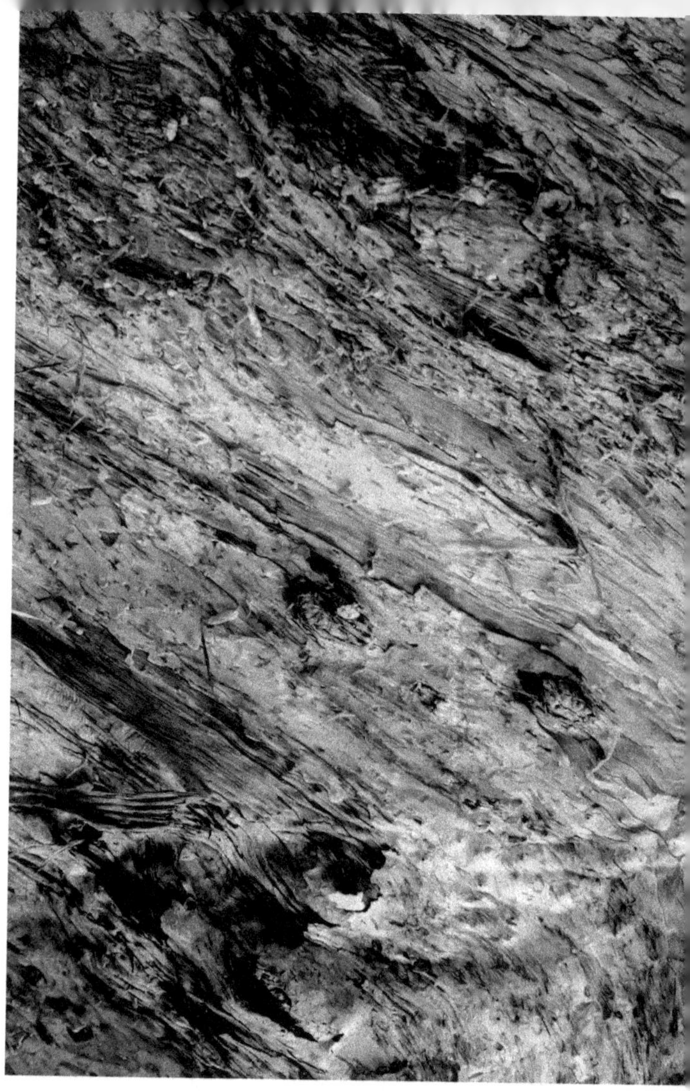

them in or if they see these commonalities. You know, there's always more than just one left. There's the left that's for social justice and government intervention, but some are also against government intervention and for a ferociously independent point of view. A lot of those in the commune movement are very similar to the Amish Christian separatist movement.

When all these things cohere, it will be around one powerful issue. John Jacobi doesn't think it's going to be climate change, maybe more globalization or economic inequality. And in a way, he's trying to figure out solutions. That's when the real danger, the tipping point comes. And it's going to come too late, it's going to be scary.

The Kaczynzki moments, they come back. There is that first shock and then there's the

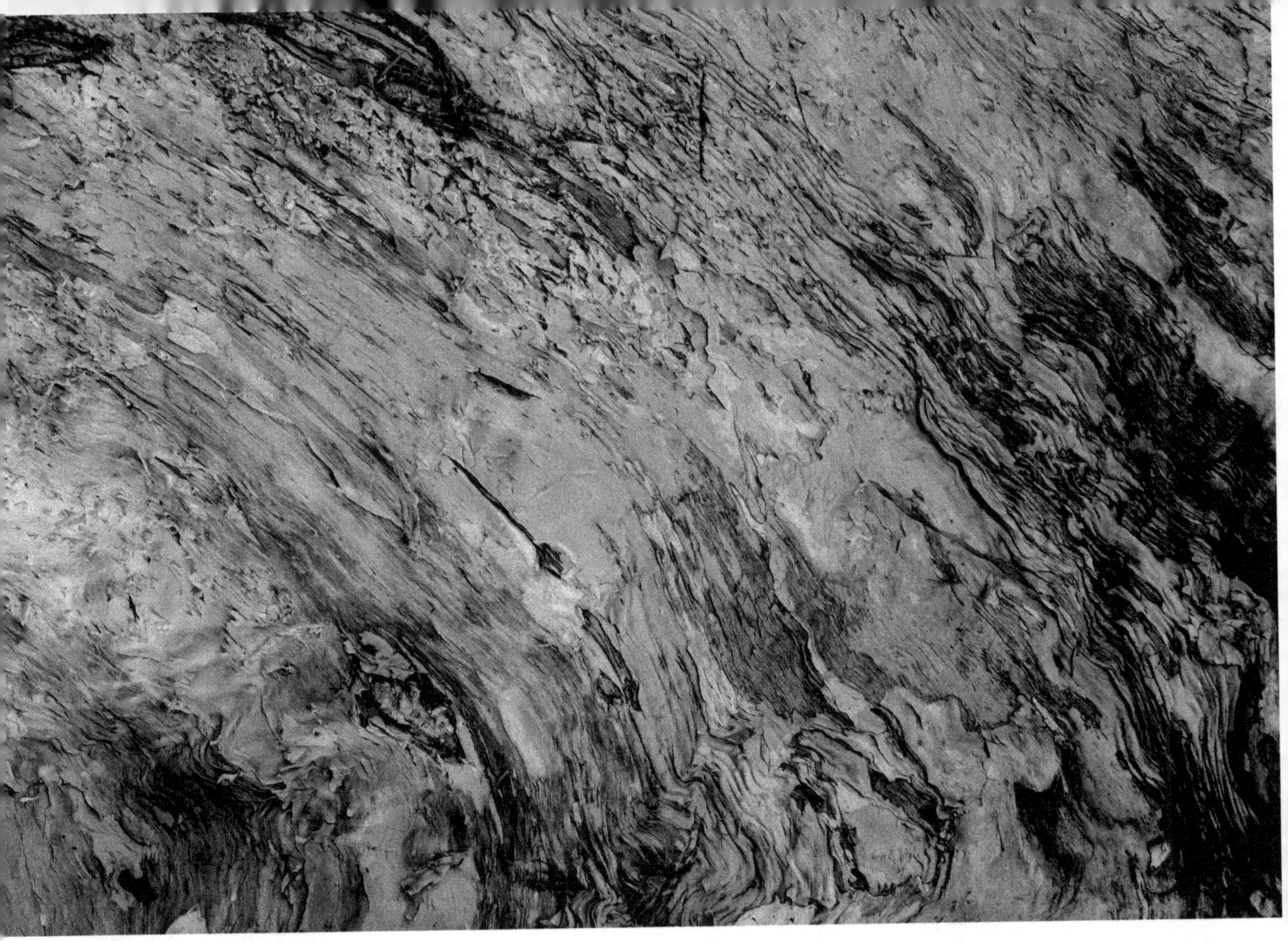

refresher course. I think about my kids growing up in this world, if they don't find some kind of magic bullet solution. It might happen. I know something about some of the science that's being done and maybe some of that will work—carbon-eating microbes at Harvard. It could happen. In Ted's terms, that wouldn't make any difference. We'd still be reduced to cogs in a system that we have no control over, our agency would be robbed, we'd become spiritual serfs. That's the Christian separatist argument.

I like John Jacobi and Paul Kinsgnorth and all these guys (leaving aside ITS/Wild Reaction who are just a nihilist path). I don't have any solutions. I don't know any smart thing to say about it. I just can't look away.

John explored the extreme edges of resistance in writing that article and put himself in positions that allowed him to directly witness the range of responses to the deepening crisis industrial civilization presents. As he said, we will see more of these "wild reactions" as we come up against more ecological limits. At least in this regard, though, reactions of this nature come from a place of deep misanthropy and do not liberate us.

I suppose that's not the point of any of it, huh?

My goals are very different in this work, while the underlying premises I work with are similar to these more eco-extremist views. The paths I wish to travel are quite different, and I hope more people will engage with their despair in less human-hating ways. To me, those paths are a dead end.

JASPER BERNES

In furthering this critique of capitalism and the industrial response to the ecological and climate crises, we must address the Green New Deal. It has been framed as a "World War II level response" to climate change. If enacted, it would (on paper) transition the U.S. economy and energy production over to a more sustainable technological infrastructure (wind, solar, biofuels, etc.), provide millions of jobs, and raise quality of life measures across the nation.

An important parallel to note here is the Green New Deal is similar to the original New Deal of the 1930s: it is a compromise. What is actually required of us is far more radical: de-industrialization, not more industrialization. It would require the abolition of capitalism, not its reformation.

We are detached from the ways our lifestyle is generated for us. We often can't account for the ways energy is produced or what's actually required to get the products we consume. Currently, fossil fuels use is absolutely integral in maintaining the vast distribution networks required to keep this living arrangement going, but sophisticated public relation campaigns attempt to sell us the idea that this can be replaced through so-called sustainable or green technologies, as the Green New Deal demands.

There are hidden costs, costs that are hidden until we choose to look at them. To get the minerals required to make solar panels and wind turbines, extremely exploitative and ecologically destructive practices have to be employed, something that cannot be solved through technological innovation.

As Jasper Bernes frames it, to produce these "green" technologies requires "sacrifice zones," regions where very specific minerals are extracted to produce the materials required to build solar panels, wind turbines, and the batteries to store this energy. This process is extremely toxic and wasteful. It may not always be stated explicitly, but these solutions are framed within the paradigm of saving the capitalist system, very similar to how the New Deal in the 1930s was an attempt to pull the United States out of the Great Depression through Keynesian economic reforms.

As Jasper explains in his article published in *Commune* magazine, "Between the Devil and the Green New Deal," what's actually required to produce these alternative energy sources is very much in line with the level of exploitation that got us into this crisis to begin with.

PATRICK FARNSWORTH: I would like to talk about your piece that was published in the second issue of *Commune*, "Between the Devil and the Green New Deal." What is the Green New Deal? What is being proposed? What are some of the programs, what essentially is its spirit? Could you go over that?

JASPER BERNES: Well, it's a bit vague, for one, and the term has been kicked around for a long time. It really depends on which Green New Deal you're talking about, because there's a broader sense of what it means. The Green Party has had a Green New Deal platform for a long time. At the same time, there is a fairly vague bill that Alexandria Ocasio-Cortez introduced into the legislature that lays out a vague project along the basic lines of a Green New Deal: to take us to 100 percent renewable energy generation within 10 years.

That means electricity from power plants would come from renewable sources or would be offset in some way. There would be no coal burning power plants, or natural gas burning power plants. So there would be free electricity, there would be no generation of emissions.

But that's only one part of it. We've got gas-burning vehicles. There's an attempt to replace the fleet of gas-burning vehicles with electrics or renewable powered cars within 10 years as well. Also, to establish more public transportation, and one of the programs is to retrofit all buildings to make them energy efficient so you're losing less energy. There are also provisions around universal healthcare and free college education.

Essentially, it fuses a fairly maximalist project of energy transition to what you might call a Social Democratic platform around universal healthcare and free education. There's also a jobs guarantee to generate full employment, provide a living wage to every single person.

PATRICK FARNSWORTH: You mentioned Alexandria Ocasio-Cortez. I know she's been the most vocal proponent of the Green New Deal as of late. She presented something to the House to see if it could get passed, but it didn't.

I remember this was kind of viral, because she brings up a lot of great points, I'm not endorsing her by any means, but I do see her as putting the Republicans under a fire, prodding them, pointing out the obvious flaws in their ideas. I'm all on board with that.

I remember there was a fiery speech she gave where she's talking about how the Green New Deal is a quality-of-life issue. You ask the kids of Flint, Michigan, whether or not they have access to clean water, and if this matters to them: this is about more than just fighting climate change or mitigating the worst of it, adapting to it and getting to sustainability.

What I sensed from that speech was it's also about addressing systemic failures in the infrastructure where water is basically unclean. There are many other problems in the country right now: in parts of Michigan and other parts of the United States, the infrastructure is failing and people's health is suffering as a result. Is that a big part of the Green New Deal, or is that something she threw in for good measure?

JASPER BERNES: Absolutely. It proposes to entirely remake the infrastructure of the United States. Introducing high speed rail, remaking the electric grid, redoing sewers, rebuilding or retrofitting every house. Also there is an access to housing provision within it.

It does also speak to the misery of life in the United States right now, which has gotten so abysmally bad. The decaying infrastructure, the high cost of housing, the high cost of health care and the general inadequacy of that health care. There's not a single person around me who isn't affected by one of those things. And it's not just about doing some abstract thing for the earth. It's about our survival and our ability to flourish. As our tagline at *Commune* has it, "a life worth living."

That's what the Green New Deal promises. That's really important. And it offers people certain optimism, which I think is a necessary basis for politics. But I'd like to take that optimism and direct it elsewhere.

WE NEED A RADICAL CHANGE IN THE WAY THAT WE DO ALMOST EVERYTHING. WE HAVE TO COMPLETELY TRANSFORM THE ENERGY SYSTEM AND THE WAY THAT WE LIVE IF WE'RE GOING TO SURVIVE THE CHALLENGES THAT THAT FACE US.

PATRICK FARNSWORTH: So this all sounds great, right? It almost sounds like we're endorsing it to a certain degree. And I know that you're not, obviously, but we're saying that this is what is perceived by those in favor of something like this. They're saying: "why wouldn't we do something like this?"

In your piece, you discuss why a lot of this isn't feasible; and not just not feasible, but actually as destructive, if not even more destructive than what we currently have. You explain in your piece what's required to build solar panels, what's required to have the energy storage required for what those solar panels create. So if I could just ask, what exactly is your big opposition to the Green New Deal, and to sustainable energy in general?

JASPER BERNES: Before I answer, I just want to say what I like about the Green New Deal, and the radical nature of it. We need a radical change in the way that we do almost everything. We have to completely transform the energy system and the way that we live if we're going to survive the challenges that that face us.

My problem with it is that I don't think it will be successful on the terms it claims. That's what matters. It needs to transform the nature of the energy system so that it's not endangering life on the planet.

It's not only that it will be just as destructive. One of the problems with renewable energy within capitalism is that capitalism has an ever growing need for more energy. You can look at the rate of energy consumption and the rate of growth of capitalist economies, and they are more or less correlated. If you bring a lot of renewable energy capacity online, that doesn't necessarily mean you're going to choke out the carbon-emitting and polluting energy. That may continue increasing.

You have to fundamentally deal with the problem of energy consumption, which is incredibly wasteful. There's no possible solution to this problem that doesn't reduce the amount of energy that people are consuming. That's a technological problem. We simply can't add enough renewable energy sources to allow everybody to continue living the way that they're living and have more people on the planet doing that. It's just not sustainable.

The good news is that we use energy in such incredibly profligate and silly ways that we could reduce this consumption. But we can't do that within capitalism, because capitalism requires people to consume more and more, and requires factories to produce more and use more energy. There just isn't a technological support for that continued accelerating use of materials and energy sources.

PATRICK FARNSWORTH: So what goes into building solar panels, for instance? What is so destructive about that? Many people feel removed from the way things are produced, especially if we're living in the United States. We don't really see where our clothes come from, or where oil comes from unless you're living next to a fracking operation.

JASPER BERNES: It really depends on what kind of solar panel you're making. There are numerous different processes, but all of them require a mix of exotic and prosaic metals mined only in specific places on the planet. That's not always because the minerals are rare. It's because these mines are so environmentally destructive that most people don't want a mine anywhere near them. So they come from places where the country is poor enough, or that the people are oppressed enough, where someone decided they'll just ruin everything by putting a mine there.

These are sacrifice zones, places that are rendered toxic so the rest of the world can benefit from the resources coming from there. It takes a lot of metal to make a solar panel and it takes a lot of energy to mine those metals and to refine them, too. Oftentimes that energy used is hard to make renewable, because it's gasoline and you have to replace it with a kind of gas substitute like biofuel, which takes many acres to produce.

You can't simply take construction and mining equipment and make it run on electricity. It's a very difficult technical problem that we're not very close to solving. That's going to cut into the benefits of going renewable. Then you're going to have to ship that stuff all the way across the planet, and then you have to replace it. A solar panel only lasts for a certain number of years: some of them about 10 years, some 15 years. You're going to have replace them. And if you install them incorrectly, that's just waste.

You have to really think about that calculation, and the same is true for wind turbines. They require a lot of resources and investment upfront, and they take up a lot more space than an oil refinery.

PATRICK FARNSWORTH: One of the main arguments I've heard from those proposing these technologies is that eventually it'll get better. We'll have more efficient solar panels that'll be able to produce more energy, and we'll have batteries that will be able to store more of that energy. We'll be able to get away from these problems eventually. What do you say to that?

JASPER BERNES: I think that's completely ludicrous, especially regarding storage. Some of the problems with renewables is that they are not continuous, they're not consistent, and they're not scalable (or what energy systems people call dispatchable).

With a coal burning plant, if you need to generate more electricity you just shovel more coal into it. But you cannot make the sun shine more, or the wind blow more. You don't have the same abstract capacity to generate more energy if you need it. Sometimes the wind isn't blowing for half of the day, and then half of the time on the planet the sun isn't shining, and often not in particular places. So you have to deal with those problems.

Now, you could store energy in batteries, or you could send it halfway around the world. But if you start adding in the technology for storage or transmission, then you're talking about an incredible increase in the use of resources. It will then take even longer to become carbon zero, because you're using so many resources. That's where it really becomes impossible—not only impossibly expensive within capitalism, but also technically impossible. You're using so many resources to get energy, and using so much energy to get those resources that may not really be of benefit.

I'm not saying all of this stuff will be useless. We're going to need to transform the energy system, and we'll need to use renewables, and we'll need some methods for storage. But it is simply wishful thinking that we're going to be able to transmit and store our way out of this problem.

PATRICK FARNSWORTH: Well, let me ask a technical question: there's a big difference between net zero and zero, right? Could you explain that a little bit? What does net zero mean?

JASPER BERNES: Net zero means assumes there are offsets that can bring you to zero. If I switch over to a solar panel, it took a certain amount of carbon to build that solar panel. All of the mining and shipping put a certain amount of CO_2 and other greenhouse gases into the atmosphere. So, I count up. And because that energy capacity is now displacing coal energy, which would generate more CO_2, I assume I get to zero once the amount of CO_2 I would have generated from my coal plant has hit the amount that it cost to make my renewable energy.

So perhaps it takes a year to pay back the CO_2 that I used to build a solar panel, because in that one year, I would have generated as much CO_2 if I used energy from coal plant as I did from building that solar panel. So I consider it "net zero" after one year. Does that makes sense?

PATRICK FARNSWORTH: Yeah. The way I previously thought about it was there could still be growth within a net zero model, because however much energy is generated from coal or other fossil fuels, you could then offset it with these sustainable technologies. But you wouldn't even be able to have these sustainable technologies without using fossil fuels to extract the materials and refine them into what is required.

Corporations and various nonprofits tend to obscure this reality, because everybody thinks we can switch to these sustainable technologies and fossil fuels will not be required. But we're still going to use these fossil fuels and only add sustainable energy on top of that to offset. So we're still going to be emitting carbon, polluting rivers and lakes. I think it's disingenuous to present it as truly sustainable, because it really isn't.

JASPER BERNES: Yeah. A lot of it is branding. I was walking in my neighborhood the other day and saw a car that described itself as a "partial zero emissions hybrid." What does a partial zero mean? You multiply by zero, you get zero. There's no partial zero. Zero is always total.

Realistically, we're probably never going to stop generating some CO_2. It's generated not just through energy production and us, but in other ways. It's released by materials, and we're not going to get around it. We're always going to be releasing some of it. The goal is to reduce it to a very low level. In any transition, we're using the energy sources that we have now to move to a different system. But the question is, which system?

Some of this math doesn't work out. It's not even net zero, because you're simply exporting emissions to other places in the world. If you ban oil in the United States, that may have an effect on the oil market that makes it cheaper, because suddenly people have all this excess oil that they're trying to get rid of. Then the oil price crashes, because the United States, the biggest consumer, has banned it. You may reduce energy use in one place, but then do something that actually makes it increase in another. It's incredibly complicated to predict these kinds of effects. But you have to think about them.

You see what I mean when I say these are dead ends, right? The paths I've highlighted are either spiritually dead and uphold a misanthropic viewpoint, or sabotage our efforts to meet the aims of capitalists, further exploiting the planet in order to perpetuate this life-destroying system. Each of these dead ends embodies extreme opposites in our potential responses to these crises. As Mark Fisher famously stated, "it is easier to imagine an end to the world than an end to capitalism." The world may end, but capitalists the world over will make a killing on the way down. They have little incentive to give up their power, unless we present them with an option they can't refuse. But wild, outrageous, nihilistic violence certainly won't cut it, either. If our activism doesn't place a love of life at its center, then it is just as life-destroying as this system.

PAUL BECKWITH

I'd like to conclude this section with some nuance, as I think there are perspectives that acknowledge all the points above. This perspective comes from physicist, engineer, and popular science communicator Paul Beckwith. Paul is well known for his approachable and comprehensive analysis of the global climate crisis through his popular YouTube channel, where he provides a multidisciplinary examination into climate science data and research through his deep understanding of meteorology, oceanography and the Earth sciences in general.

Paul certainly understands the predicament we are in. He understands the alarming levels of warming we are witnessing in the Arctic region, and how this is producing a dramatic loss of sea ice cover, diminishing the albedo effect (reflection of the sun's heat) in the region, leading to what he describes as a "blue ocean event" (an ice-free Arctic during the warm season). He has explained extensively how these rapid shifts in the global climate system are leading to more abrupt and severe floods, droughts, wildfires, and storms across the planet. The implications are vast and dire, in particular when it comes to global food production and distribution.

As Paul states, this may well just be something of an engineering problem. He has certain views on what can be done in light of these abrupt changes, including various geoengineering schemes to cool the Arctic region, and other techniques that may give humankind the time needed to avoid the worst impacts of abrupt climate disruption. I contend with Paul's views on this subject, and I pose some objections I have, including the unintended consequences that would likely result from these attempts. Also, I express what these geoengineering schemes may end up serving within our current infinite-growth economic paradigm, which he is quite aware of.

In spite of my differences with this view, I felt it's important to present a more nuanced view. I tend to get locked into rigid points of view on these subjects, and having someone like Paul as part of the discussion forces me to contemplate these possibilities.

PATRICK FARNSWORTH: By no means am I a scientist, so I want to pose this very humbly. You said, it's an engineering problem. But from what I gather, we see how these disruptions are a result of human industrial activity.

You mentioned pesticides, how those are impacting insects and pollinators. And there are various other ways human beings are impacting the life systems of this planet, the climate system, everything. It seems to me the problem is that we were to try to do those things you said.

Basically, when you create something that'll help save the Arctic from completely melting in the warm season, or you're trying to alter oceans in the ways that you've mentioned, there may be unintended consequences. It would be really difficult to anticipate what those consequences would be, similar to the way humans use a technology like dams. We don't necessarily anticipate, or maybe the engineers don't seem to care, how the dams would impact the river ecology. For instance, the way salmon migrate and breed, and how disrupting that impacts the entire ecosystem, and not just nonhuman life, but human communities as well.

It seems to me that there are all kinds of unintended consequences from these mass engineering projects. I just want to pose that, trying to do these things, which I think could be very noble, may make things worse. Perhaps a certain feedback loop, or certain tipping points, or certain other things that we wouldn't have anticipated may result.

PAUL BECKWITH: Right. Yeah. So, unintended consequences. We're talking about the risk here, really.

You brought up that there are risks, which is a very valid point. There are risks with doing some of these enormous scaled-up projects, to try to remove CO_2 from the atmosphere or to cool the Arctic, and so on. But we can't look at those risks in isolation. We have to look at those risks and weigh them against the risks of not doing anything. Or weigh them against how our climate system is changing right now.

Our global food supply is under threat. I mean, our civilization is under threat, right? Our cities are being destroyed. Huge states are being inundated with water. These are unintended consequences of burning fossil fuels continuously.

Surely, there are unintended consequences of putting sulfur dioxide up into the atmosphere to cool the planet. If it starts having unintended consequences, then we could stop doing it. It's not like fossil fuels, which we can't just stop. We're not able to stop our addiction to fossil fuels. I would argue that the risks are enormous from doing nothing, and they far exceed any of these risks that I can think of for trying to. Let's do some of the more innocuous things. I mean, let's plant a trillion trees on the planet to soak up CO_2. I mean, what are the unintended risks of that?

THERE'S THIS PUBLIC PERCEPTION THAT WE SCREWED THINGS UP SO BADLY ANYTHING WE DO TO TRY TO STOP THE PROBLEM IS ACTUALLY GOING TO MAKE IT WORSE. WE END UP GETTING CAUGHT IN THESE THINKING LOOPS, AND IT STOPS US TAKING ANY ACTION AT ALL.

Let's start projects to try to do things like that, or some marine cloud brightening, which is just spraying seawater into the air. Let's fit ships with that, and start doing that in the Arctic and see if we can start cooling the Arctic. There is a company, a start up called ICE 9/11 or something, and they want to use these very small micro beads, glass beads, basically silicon dioxide, to make the ice and seawater more reflective. Putting glass up into the Arctic, small microbeads, that would be horrible for the environment when you first think about it. But then when you realize that one of the most common types of plankton in the Arctic are diatoms, and they form a glass skeleton, a very tiny micro skeleton that is part of their body when they grow. So we're talking about just trying to do some things that nature is already doing. These little glass beads proliferate in the environment in this type of plankton. So the idea is to use these things on the ice, to try to make the ice more reflective.

We have nowhere near the number of people that we need on the planet to think about and to trial these things, because there's this public perception that we screwed things up so badly anything we do to try to stop the problem is actually going to make it worse. We end up getting caught in these thinking loops, and it stops us taking any action at all.

You know what? I really do not understand the "doomer" community who says we're all gone in 10 years. And then you talk to them about sulfur aerosol injection, and they say, well, that will make things much worse. Unintended consequences. And I say, well, what do you mean? Do you mean that will reduce the ten years that we have to eight years? Where's your logic on this?

PATRICK FARNSWORTH: Right. [both laughing].

PAUL BECKWITH: What do they actually mean? And they never you know, they can never answer that question. I know it may not work. I mean, we may not be able to do anything.

PATRICK FARNSWORTH: Yeah, right.

PAUL BECKWITH: But we haven't tried. We haven't put any resources into it. What happens when we stop subsidizing fossil fuels, when some of that money goes to subsidizing some of these other things—which is what the students are demanding, and Extinction Rebellion? All of these groups are trying to raise awareness on how serious the climate problem is.

PATRICK FARNSWORTH: This question has a really strong political dimension to it. One of the trends I'm seeing with these youth climate marches and this massive effort from nonprofit organizations? As certain journalists have described it, the nonprofit industrial complex has really deep ties to the very wealthy class of individuals that want to transform our economy into a so-called sustainable economy. But it's still in a capitalist economy. It's still based on the idea of growth.

That's one of the major underlying flaws of many of these major projects and big activist groups: they're not really critically analyzing the very systems that have gotten us into this situation to begin with. One of the fears that I have is we will start to do things like you're recommending, and that will be only in service to maintaining a socioeconomic system that does not blend in any real way with actual sustainability. It's very possible that the narrative is currently being hijacked by these individuals and these groups.

PAUL BECKWITH: Yes, right.

PATRICK FARNSWORTH: That is a big fear when I think about your geoengineering suggestions. I'm like, okay, let's figure this out. Let's see if we can throw a few ideas out there. And I don't want to sound defeatist and I'm really trying...

PAUL BECKWITH: I know exactly what you're saying. Let me give a good example that fits in with what you're saying.

In the 70s, the Club of Rome published their famous book, *The Limits to Growth*. They said that the human population couldn't keep growing, and I guess the population was probably not quite 2 billion when that book came out. They said that the food supply was increasing at a certain rate, while population was increasing at an exponential rate, and that alone would put an upper limit to population. There were limits to that growth. They also talked about minerals and other things that people use.

And what happened, of course, was that we had these enormous breakthroughs in farming and we managed to double yields, double the amount of crops that could come off a field and then double it again and double it again. Yields went way, way up. That avoided the upper limit at that time.

So let's imagine that technology had not come along. If that technology had not come along, there would have been massive starvation, because population probably still would have grown until there were massive food shortages. Massive starvation would have put a limit on population, and humans would have had to actually sit up and say, okay, we have to take some hard actions here. Do we want to have these continuous cycles where we get mass die-offs of humans from starvation, and then society continues and population grows exponentially again, and then another mass die off? Or do we want to talk about the whole

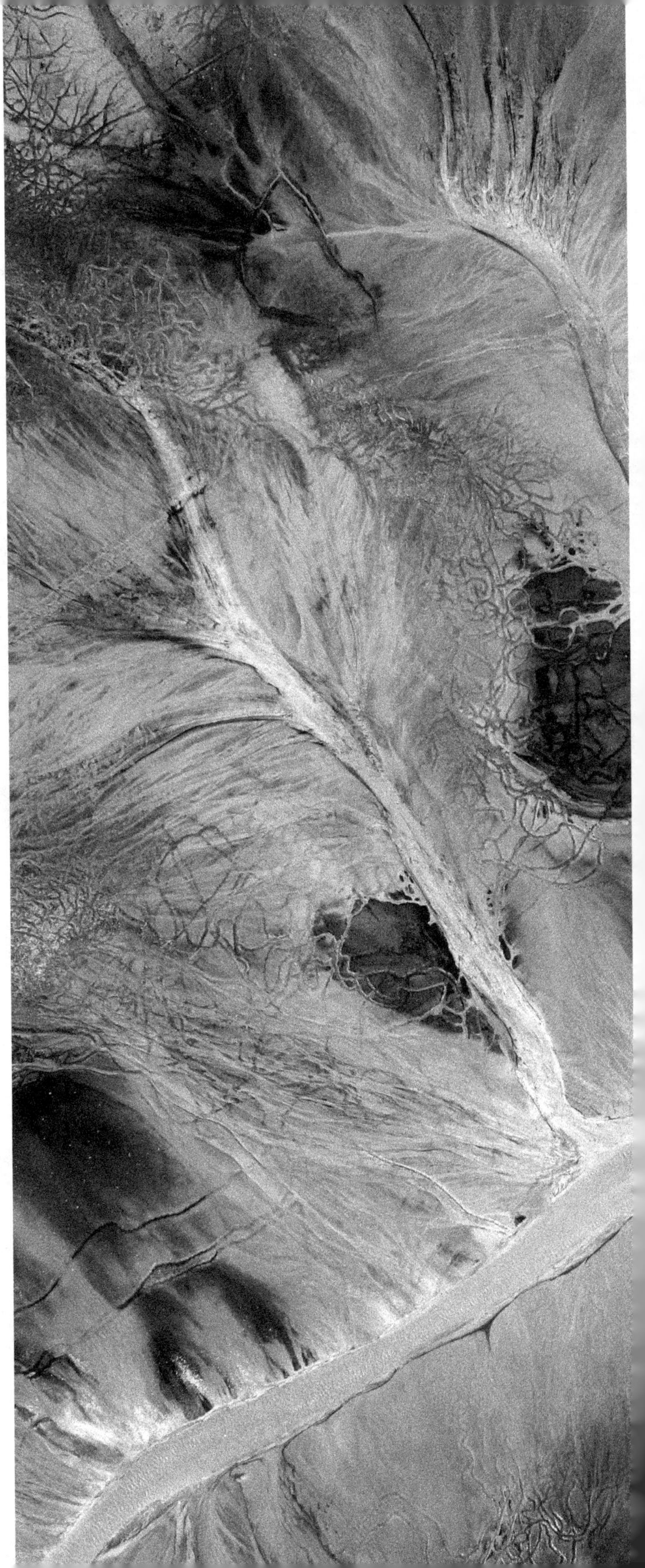

system of the planet and how many people can live on the planet and figure out how to live within the means and be truly sustainable?

And if that had happened, we wouldn't be having the climate crisis that we're having now. We would have a global population of a couple billion and not 7 and a half billion. We would have had to deal with exponential growth, consumer growth, and stuff like that. And we'd have quite a different world.

So I see exactly what you're saying. Geo-engineering could be like these huge improvements in food production back in the 70s. These things could work for humanity in the short run and allow us to continue our way of life. Capitalism can continue, population growth can continue, all until we reach another wall we eventually can't solve. And then we have a population crash.

When I talk about these things, I talk about them from a survival standpoint. How can we preserve some semblance of humanity? How do we restore an ecosystem? How do we ensure that we don't have a mass extinction on the planet that includes ourselves? These are emergency stopgap measures, but they're no good unless we address these underlying issues, like population, rampant consumerism, and all of this other stuff.

Just in the oceanography I've been looking at, there's a great section on the emirate of Dubai and how they're changing the coastlines, how they've basically dredged up millions of tons of sand and created artificial islands in the shape of a pineapple.

PATRICK FARNSWORTH: Yeah. [laughing].

PAUL BECKWITH: They actually have all of these very wealthy people that are buying places on these artificial islands and these things are all sand, you know? Don't build houses on sand. The excesses of humanity—it makes you sort of wonder, do we even deserve to have this planet? [laughing]

PATRICK FARNSWORTH: Yeah.

PAUL BECKWITH: It's crazy. And then it extends to politics. These ideas of removing CO_2, and cooling the Arctic, they're not for preserving any of the systems that we have, as far as rampant capitalism and exponential population growth. It's more of a survival thing, so people can live longer-term on this planet. So that everything doesn't completely collapse for us, which is what it's doing. That's what exponential growth does.

PATRICK FARNSWORTH: Definitely.

PAUL BECKWITH: I don't know if that answers your question, but I guess maybe the pain needs to become greater for people before they'll sit up and ask: "what are we going to do?"

The pain will get greater. And really, what can we do? Whether through the engineering schemes posed by Paul or not, our response to this unfolding global catastrophe must be grounded in our deep and abiding love for the Earth. Wild, unrestrained reaction against our brothers and sisters is not grounded in love. Allowing our energies and activism to be co-opted by capitalist interests is certainly not, either.

I've tried to explore the edges of a very wide spectrum of potential responses I view are dead ends, and in service to goals I deem to be a waste of our collective and individual efforts. Considering our time here is short, what does a sacred response look like?

PART SIX
WE WILL NOT ARRIVE INTACT

THE FUTURE IS UNWRITTEN.

How we choose to respond matters. Even in the face of what lies ahead, our response to these times of trouble may be varied, but it will always be layered with grief. This cannot be avoided, and part of what I call our sacred response will come from this place of grief. As such, we shouldn't give into nihilism and cynicism, and if I could be dogmatic about it, our actions should not be based on an anticipated outcome, that we can be saved or that there is a battle to be won (to "save the Earth"—which is just a coded phrase for "save this civilization"). We are past that point.

Regardless of what is coming, we must demonstrate our love for the Earth, in all the forms that comes in. Our love is not an abstract, it's as tangible and as real the air we breathe and the soil beneath our feet. How we choose to be in this time is as important, if not more so, than any other time previous. Our love does not recede, and is not diminished, because death is around the bend. This is as true for the individual as it is for the collective. Our acceptance of limits is what makes our lives all the more meaningful, and any system that infringes on our ability to fulfill our obligations as members of this planet's community of life should be resisted, full stop.

In the previous section, I pointed to a range of responses that, at least in my view, will not elicit anything close to the kinds of actions that are actually required of us. Whether it's the vengeful, murderous, and wild outbursts of violence of misanthropes and nihilists, or the concerted efforts to remake capitalism under a "sustainable" energy paradigm that only deepens and further replicates the worst aspects of this world-eating machine, these responses will not solve, let alone resolve, a damn thing.

STEPHEN JENKINSON

To know how to proceed, asking the right questions is crucial. Grief is the only sober response as this globe-spanning catastrophe plays itself out. Whether we choose to accept grief into our lives, our communities, our activism, our expressions of love, will make all the difference in how we proceed. We can't even ask the right questions without accepting even this most basic of things.

We must wonder why we can't seem to ask those right questions, or why those questions rarely arise. Part of the reason may be our lack of elders. In the view of the next person I would like to feature, living in an environment bereft of elders has denied us something deeply crucial. As Stephen Jenkinson elaborates, the elders that would emerge in our time would not be those of any other time previous. Our time demands something quite different.

Stephen authored two books relevant to this subject. The first is *Die Wise: A Manifesto for Sanity and Soul*, which gives profound insights into our death-phobic culture drawn from his countless encounters with death and dying as program director for a major Canadian hospital palliative care unit. The second is *Come of Age: The Case for Elderhood in a Time of Trouble*, in which Stephen argues that elderhood is a function rather than an identity—it is not a position earned simply by the number of years on the planet or the title parent or grandparent.

THIS QUESTION ABOUT ELDERHOOD COMES FROM ITS ABSENCE.

Why is it that the dominant culture of North America does not produce the conditions necessary for elderhood to flourish? Our denial of limits, of death, is directly tied to this question. In my interview with him, Stephen discusses the historical and cultural conditions that have led to this unexplored and unexamined crisis, and speaks to what elderhood in our time could look like, framing it within the profound realization that elderhood can only flourish when the appetite exists for it.

> I am making the case for elderhood, not for easy agedness. I'm doing so mostly by wondering what happened. Because something happened. Something happened to ancestors and elders and honour. There's work to be

> done, and there's an old wisdom to be learned where there used to be the wisdom of old, and you can't fix what you don't understand. That's where we're headed: to grievous wisdom. Let us see if we can bear the sound, the particular sound, of no hand clapping.

In our times of trouble, where are the elders?

PATRICK FARNSWORTH: You've pointed out that we have many elderly people in our society, but we lack elders. What is elderhood in your perspective? And why are we in such desperate need of elders in this time of trouble?

STEPHEN JENKINSON: We can actually work backwards here, from some very observable things, rather than asking, "what is elderhood, what does it look like?" You might wonder, "what doesn't it look like? What would be a sure sign that elderhood is in some kind of atrophy?" This takes some imagination, but it also takes more observation.

Popular rancor today is fixated upon several things in North America. One of those is identity politics. This doesn't personally concern me, but I'm paying attention to the current fascination with identity, because to me it is a smoking gun. This is one of the things that you begin to infer backwards from rather than going with the idea that inherently being tormented, tortured, and precise about one's identity is a good fix. It might be an attribute of the thing you're trying to fix instead. The lunge towards ever greater fixity in identity seems to suggest a fundamental adriftness regarding what one "means." Not our intentions, but the meaning of our very existence.

I SUSPECT THERE WAS A TIME WHEN "RESPECT YOUR ELDERS" WAS NOT PRESCRIPTION.

This meaning is somehow up for grabs every time identity clamors for attention, and it does so in a way that's not very dignified. I'm not sure it trades in anything substantial, other than feelings, a sense of how I feel. And when feelings become your map, you're in for heavy weather, because that's what the nature of feelings are. They're weather. They blow in, they blow out.

Where I'm sitting today talking to you, it's very sunny. If I were an architect and could build a house in a day, and if I was preoccupied with the feeling of this particular day, I would build a Phillip Johnson house—nothing but glass, virtually no walls to speak of. And the truth is, within four months here, it would be at minus 30 Fahrenheit. So that house will be of absolutely no use to anyone. It won't even be a greenhouse. It'll be a disaster.

This is what I mean by predicating your entire self-project on how you're feeling today about something. I'm not saying there's no such thing as weather. I'm simply saying that it changes: it blows in and blows out. To extend that understanding to the question you've asked about elderhood, I might come up with something like this.

This question about elderhood comes from its absence. Nobody would ask about it if we were awash in elderhood. Nobody would be wondering about it. Not that many years ago, we were awash in death phobia. I squeaked a little book called *Die Wise* and then people started asking me about it, but I was never asked about it earlier, ever. By the same token it seems the concern over elderhood speaks to its virtual nonexistence among us.

More importantly: I have scores of older people coming to the things that I do. And what they're hoping for is that their elderhood will be recognized, that it will be rewarded, it will be employed, but most emphatically they look for an imprimatur for their incipient elderhood, that it will burst forth in all its raging glory. If there were such a thing, it certainly wouldn't need me to confirm it, because who am I but another person struggling along?

If older people are floating from one retreat center to the next self-help event and so on—they're the people that basically keep the lights on in places like that—is this a sign that elderhood is so rife that we have to elbow older people out of the way to get to the trough of elderhood? You see what I'm saying? In my experiences, it's quite the contrary. When you have older people looking to be identified as elders, this is as much of a sign of the absence of elderhood as anything could be.

In North America, we are an ever-aging population. If age meant elderhood, you can do the math immediately. We're an ever more elder-ing culture, if you want to put it in that awkward fashion, meaning: with our accumulation of age comes our accumulation of aged people. With that comes the accumulation of life wisdom. And with that comes the accumulation of elderhood. That's the automatic slippery slope that everyone

assumes, because how could you come to another conclusion if you don't question the idea that trauma experience automatically translates to wisdom at some point?

I'm suggesting to you that it does not and it never has. I was raised with the following declaration from a young age: "respect your elders." This was told to me sometimes as a correction on my behavior as a young kid, and more often than not there was utterly no explanation that accompanied it. None whatsoever: no elaboration, no justification. It was nailed to the sky for all to see. One respects one's elders.

So let me wonder about this for a minute. It seems self-evident, and I'm not saying it is not advisable nor a deeply meritorious way to behave. But I'm wondering about the kind of mandate that's carried in the phrase. I suspect there was a time when "respect your elders" was not prescription, it was description. When people said "respect your elders," they were trading on the understanding that elders were imminently and irretrievably respectable. Not that they just earned it, but that it was self-evident that they had earned it. So when people were trafficking in that phrase, they were acknowledging the obvious, as well as advocating for it.

Now, here's the dilemma. When the question of elderhood comes, an awful lot of people over 60 years old basically raise their hand not to volunteer, but to get their allotment. But what were they doing in their peak income-generating years? What was the world like in the dominant culture in North America? What was going on? I won't trouble you with historical details, but it's recent history. I could mention the word Monsanto, and that would cover a lot of territory because Monsanto occurred during the time frame I'm talking about. And of course the maniacal charge towards climate change was accelerating during this time, though the science was already clear.

So while the case was being made that something fundamental had to change, the people who had their fingers on the trigger, had their hand on the steering wheel, had their arm at the helm—all of these people of the generation I'm describing now were in positions where they could have acted on these troubles. I don't say fix them—I'm saying they had the opportunity to behave and conduct themselves as if we were about to enter the shit.

So my question is, what did they do during that time? I'm sad to report that, when these people who now want to be recognized as elders had the chance to be elders-in-training at the beginning of their middle age, they appear not to have done so. The world that people like you are about to inherit from people like me bears the marks of truancy in the generation I'm describing. Not even mistaken notions, but abject truancy. And if any of that holds any water, then my question becomes: when somebody says respect your elders, I say, "respect what?"

PATRICK FARNSWORTH: I was definitely told as a child to respect my elders, but oftentimes the question was why or who? Obviously they meant parents and grandparents. But I think kids are really good at spotting bullshit. They may not be able to articulate it very well, but they understand that something is a little off about that. And what you just said there is really profound: that there was a time when it wasn't a matter of telling you to respect your elders, but rather saying we respect our elders because it was just implied, it was pervasive.

I sense between generations now, there is no sense of appreciation or respect for what would be considered our elders, the generation that came before us. There's resentment and frustration, anger and anxiety.

STEPHEN JENKINSON: And dismissiveness. Don't forget that one.

PATRICK FARNSWORTH: Yes, exactly. Because I think the guidance we look for isn't there. Like you said, these people are asking to be respected as elders, yet they don't hold anything that resembles elderhood. I think we intuitively sense what it means to be an elder.

In an interview, you mentioned elders you had in your life, people you knew when you were growing up. It's almost automatic that younger people can sense if an individual embodies elderhood. You sense it and you feel it and you're attracted to that person. You want to be around that person, you want to learn from that person. And that person then becomes an example. And you may not even realize it at the time, but it's oftentimes in retrospect that we really integrate the lessons that came from being around those people.

STEPHEN JENKINSON: You've said something very important that needs to be emphasized. I should preface this by saying, I know there's people who are living exceptions to the rule I've been describing. There's no question about that.

You and I are talking about the same dilemma, which is not fundamentally of attitude or misapprehension. This is fundamental to our way of life. That's the first thing. Secondly, I want to ask, apropos to what you said: where did you learn "to respect your elders," aside from being instructed to do so?

The answer you gave was that, when you were around them, there's an intuitive uptake. It's almost alchemical. It's really not by instruction at all. It's from some kind of remarkable example, and that example has consequences for you when you're very young beyond whatever it is that you were seeking. Often you can feel drawn to somebody in a way that doesn't resemble what you might have been seeking for your self-improvement or betterment. The problem is that your instinct for it is formed by and then informed by its presence. So here's the horrible possibility: if there's enough of a disappearance of elderhood, the instinct to recognize it among younger people can go into atrophy.

IT OCCURRED TO ME LONG AGO THAT IT IS NOT OLDER PEOPLE THAT CONFER ELDER STATUS UPON EACH OTHER. ELDERHOOD IS FUNDAMENTALLY RECOGNIZED BY THE PEOPLE WHO SEEK IT, NOT THE PEOPLE WHO SEEK TO BE IT.

Do you see what I'm saying? It occurred to me long ago that it is not older people that confer elder status upon each other. Elderhood is fundamentally recognized by the people who seek it, not the people who seek to be it. It is in the wheelhouse of young people to craft elders in their midst by their willingness to seek them out.

So you have this double helix arrangement between the generations, which skips the medial generation, the parental generation. It operates something like this: by virtue of their appearance, they send up a kind of tendril of longing in you for them. But their absence doesn't do that.

Years ago, I went out with an Italian woman whose mother was an insanely good cook. I heard the two of them talking one day when I was sitting in the kitchen, and her mother said to the daughter: "well, you know, food makes hunger." She said that and just left it in the air like it was the most self-evident thing in the world. I tried to make sense of it, and this is the sense that I made of it. The Cartesian cause-and-effect fascist that we walk around with in our minds all the time dictates that it's the absence of something that creates the desire for it. But you know, there is the distinct possibility that nothing of the kind actually ensues: it's the presence of something that creates the longing. This is why you can eat, and then eat a couple hours later, though you are not hungry. The food appears again, and you think to yourself, "yeah, I could do that."

WHEN YOU DECIDE THAT PEOPLE OVER A CERTAIN AGE ARE EASY TO DISCREDIT, ARE AN EASY AND PROPER TARGET FOR YOUR VENOM AND YOUR DISAFFECTION ABOUT LIFE, DO NOT BE SURPRISED THAT WHEN YOU BECOME OLD YOURSELF YOU HAVE INTERNALIZED THIS LESSON.

If you apply that understanding to the experience of elderhood, you begin to see something beyond a national tragedy. It's at the level of pandemic: if elderhood functionally disappears over more than one generation, there is a palpable rupture of continuity. At a sociological level, it may be irreversible. People know this about the disappearance of culture and language. You have the grandparental generation who was raised with the language. You have an intermediate generation, those who were there when the conquest or conversion to another religion came. This generation has both languages, but their young children are raised to benefit from the changed circumstance. The old language doesn't benefit them or serve them in this changed circumstance. So the rupture between the grandparents and the grandchildren linguistically is so severe they often can't talk to each other, even though they were technically born in the same country (but certainly not in the same culture).

To give you a feel for what this looks like, consider the Inuit of northern Canada. An old woman was interviewed one time and the interviewer asked, "what's the worst thing that's ever befallen your people?" This is a deeply indelicate thing to ask somebody from any kind of traditional culture, and it was clear to him that she didn't know. So he started suggesting things to her. In fact, what she was trying to do was politely avoid answering the question. He kept prompting her and finally she said, "I think it was television."

The interviewer couldn't believe television compared to starvation and forced relocation by the government. He said, through the interpreter, "how can you say that television is the worst thing that ever happened?" And she gestured to her grandson, who was sitting in the next room, plastered to the television, of course, and she said, "ever since this thing came, we can't talk because I can't talk TV," by which she meant English.

This is what happens not just linguistically but at the level of elder function. It can literally waste away within two generations. And you may have legions of people in your children's generation who are so adrift that they embrace self-determination as a religion, as a personal obligation, as a moral act, and as a political orientation. Because the willingness or the ability to trust in anything that came before you is at least as corrupt as the ecosphere might be by that time.

PATRICK FARNSWORTH: There's a lot of thoughts going through my head right now, I'm trying to figure how to articulate this next question because it's going to sound rather naive.

STEPHEN JENKINSON: Well, let's risk it.

PATRICK FARNSWORTH: Okay. I really want to know your perspective on why this is the case. I think this ties in to your work into the dominant death-phobic culture of North America. They're obviously connected in some way. What is wrong with our culture? Why is it like this? Why are we so adrift? What has happened to create these conditions?

STEPHEN JENKINSON: I've always found a good question to be worth 40 good answers, because a real good question survives all the answers that are flung at it. And the spirit of inquiry remains even after the question is answered. So I hope in trying to answer this that what we're both doing is lending credence to the question rather than trying to defeat it by coming up with an ever more elaborate and perhaps nihilistic answer. There are many things to point to. There's no such thing as a beginning for any of these things. But you could speak of plural beginnings and you can't isolate them in a particular moment, because there was a moment just prior to that which predisposed that moment and so forth.

A couple of things occurred to me. I'm old enough to remember the 1960s, though not old enough to really be included. By 1969, I was 16, a bit too young to wade in with both feet. Then, it was an article of faith in the politicized generation that you should "never trust anybody over 30." It was a viral notion: there was a sense that there was a contagious quality to people over a certain age, that they had been so deeply compromised by working "in the system," by participating in the system or even being educated by the system that they could never really recover their dignity or their sense of right and wrong.

It was extraordinarily naive, and frankly a kind of apartheid. You had people in their late teens through their late 20s traipsing about the cultural landscape, automatically denigrating anything or anyone older than them and their lived memory. And what was their background? It was the Second World War, of course, and the European Holocaust, the atomic bomb, and so forth. They found what preceded them so fundamentally discredited that they had no interest in engaging. Their belief was that when it was their turn everything would change, just like many women believed that once they took their places in the boardrooms of corporate America, corporate America would be humanized. We've been on that experiment for some time now, and I wonder how the humanization is working out?

By the same token, when you decide that people over a certain age are easy to discredit, are an easy and proper target for your venom and your disaffection about life, do not be surprised that when you become old yourself you have internalized this lesson—how despicable it is to grow old—that you can't renegotiate it. It's no surprise to me that the people who did walk around in that air are now looking to be venerated as elders without particularly acknowledging themselves as old.

There's a line from a very famous song of the period, called "My Generation." The line is, "I hope I die before I get old." It was an enormously popular song that is still ringing in the memory of people of a certain generation now wondering if it wouldn't have worked out better had they died before they got old.

In an older understanding of the term, you could say that any place populated by spontaneous mass migration as North America was, the cover story is that people were looking for a better life. The truth is, people were running for their lives, but they had no life to run towards. It's not like they'd had an extraordinary run of beautiful luck, gorgeous democracy and a pool in the backyard, and then there was a slight downturn, and so they headed for America. Most of these people were running from every nightmare conceivable.

The irony is that they ran the way haunted people run. That means they bring their ghosts with them. In trying to escape your haunting, you bring your ghosts. And this is what America has become. It's a ghosted place. America was a European fantasy, and as the generations distanced America from Europe, it became a haunted, adolescent place, and it still is.

That's going back three hundred or four hundred years. When people left their ancestral lands, they left behind languages, understandings of life, poetries and songs. All the fundamentals of culture became instead a hyphenated identity, and identities suddenly became very, very important. Why? Because there weren't any identities. By definition, you were not at home, and what do people who are not at home behave like? Ask any Indigenous North American and they'll answer you, because that's their recent history: being overrun by people who are not at home, ghosted, haunted, allegedly seeking freedom, but denying it to everyone they met when they got here.

The loss of any ability to live a sustained cultural continuity is bound to catch up with you. When does it catch up with you? Not when you're in your peak years of physical dexterity and strength. Not when things are going swimmingly well for you. It catches up with you when everything starts going south, whether it's personal health, your outlook on things, or your ability to believe in anything anymore and proceed accordingly.

If you put those two things together as bookends, with about 400 years separating them, you certainly begin to get an idea of why the idea of cultural inheritance became seen as something that was better left behind. If it wasn't left behind, it became a burden. And now many people your age are telling me candidly and openly that their resentment of people twice their age is enormous, largely because of old people absorbing an astronomical amount of the GNP.

It's not like they haven't already had their way through their entire lives, this older generation. And now they're demanding euthanasia as a right, aren't they? And LSD as a right, and legalized marijuana as a right. And of course, apparently they're demanding elderhood as a right, and the demands never end. And young people hear this litany of demand that's now sucking up something like 80 cents of every health care dollar, and their resentment is just around the corner. Well, it's not around the corner. It's part of the dialog now.

PATRICK FARNSWORTH: So what would elderhood look, feel, and be like in this time that we are in?

STEPHEN JENKINSON: If elderhood is an identity and swept up with all the other identity clamor of our time, then what you'd do is look for people with that "personality type." Or maybe there's an elderhood MMPI—a psychological testing apparatus whereby we can tease out elder-tending personality types. Then you could identify the particular kind of elder you're looking for and dial that in. Instead of speed dating, you could have speed elder-ing. You could just tell them what you're interested in as far as being mentored and ten older people pop up and you get to choose. I wouldn't be the least surprised that would come, and somebody is going to work on it as soon as they've heard me suggest it.

PATRICK FARNSWORTH: [laughing] Yeah.

STEPHEN JENKINSON: But I'm going to offer an alternative to that. I'm going to suggest that elderhood is not a figment of personality or an aspect of identity. It has nothing to do with the particular qualities of individuals.

Elderhood is first, foremost, and will always be, a cultural function. In that understanding, an elder is a culture worker and so is not inherently an old person. While all elders tend to be older, not all older people are elders. So if elderhood is not a personality type, what else could it be? You've said it's something to do with culture. But what precisely? Well the answer is the subtitle of my book: it's why I called it "the case for elderhood in a time of trouble." I believe the particular wrinkles of elderhood are dictated by the times in which potential elders find themselves. They were born to a particular time and the particulars of those times dictate what elderhood must be now.

So this means elders themselves must be on the steep learning curve, and they must be deep running students of their times. Their responsiveness to their times is what qualifies them. The word *responsibility* really works here. It's not a sense of burden: sense of responsibility means simply the capacity to respond, and we need to distinguish respond from react. React describes certain responses that attempt to satisfy you, or assuage or reassure you. Whereas the capacity to respond might have nothing to do with you trying to feel better about anything: instead it comes from your sense of moral, political, cultural, or spiritual obligation. It comes from fully trying to inhabit the conditions of citizenship if you will, but citizenry not to particular geopolitical identity: your deep citizenship is a devotional one, rather than from affiliation. In that sense, the work you join yourself to is dictated by your time's troubles. You're a citizen of a troubled time.

If that's possible, then it means that elders are not in the business of getting themselves recognized, they're in the business of recognizing. In a time when elderhood has gone into terrible abeyance, it becomes the elder-ing responsibility to recognize incipient elderhood in their midst, and to proceed accordingly: acknowledging it, recognizing it, corroborating it, and authorizing it without ever trying to be included in it or to benefit directly from it. You follow what I'm saying?

PATRICK FARNSWORTH: Yes.

STEPHEN JENKINSON: It's a radical re-understanding of what it means to be an elder. It's not a club you get to join. No elder in a time like this would ever call themselves an elder. Because it is the responsibility of the people around them to recognize elderhood in their midst and to corroborate it.

And if it doesn't happen, it's because there are no elders to do so. The appetite for elderhood has gone missing, because young people are not exposed to it and their appetite for it atrophied, and they have traded it for self-reliance and a principled anxiety that masquerades as having a conscience, a chronic free-floating anxiety where you care about everything but only enough to paralyze you.

This is an awful lot to say in response to a short question. More simply, the greatest elderhood skill is not the skill of knowing how to be an elder, it is the skill of knowing how to have elders in your midst.

DR. KARLA TAIT

In the same way our appetite for elderhood can atrophy when it loses its currency in the culture, so too can our connection to the land on which we stand, and in turn the whole living planet. If we haven't had this connection, if that relationship hasn't been felt and present in our lived experience, it's almost impossible to imagine that we would remember ever having it. And it's even more difficult to imagine we would care to even defend what's left of this living planet if it felt separate from us, an object to be used to serve the interests of the dominate culture. We should now know where this logic will lead, and has already led, us: over the edge of a spiritual and ecological abyss.

Part of the sacred work required right now is to deconstruct this disconnection, as inheritors of this settler-colonial culture, with the lands we stand on as dominators, as exploiters, and abusers, rather than standing in relationship to it. I believe we can trace the roots of this ecological and climate crisis to this separation. As so many other people I've interviewed for this project have suggested, this disconnection stems from a deep trauma, one perpetuated generationally in the wake of the severe ruptures that settlers were attempting to escape from in their homelands in Europe (as expounded upon by Silvia Federici). The problem, of course, is those traumas—born from hundreds of years of violence—were never resolved, and inevitably laid the groundwork for one the greatest crimes to ever occur in human history.

To this day, the Indigenous inhabitants of this land (I'm speaking as a North American here) struggle not only to demand autonomy as First Nation peoples within the borders of colonialist nation-states (e.g., the United States, Canada), but more specifically to protect their lands against the intrusions and violations of the fossil fuel industry. These intrusions are fully backed by the State, as we have witnessed in the United States with the proposed construction of the Dakota Access Pipeline in 2016–2017, and over the past decade in British Columbia with the Wet'suwet'en and Gitxsan First Nations' resistance to the proposed Coastal GasLink pipeline project.

This next interview, recorded in January of 2019, is with Dr. Karla Tait, member of the Wet'suwet'en First Nation and Director of Programming for the Unist'ot'en Healing Center. She provides more context on the ongoing confrontation with the Canadian State and the pipeline project that threatens to push its way through their unceded territory. She recounts the invasion by the Royal Canadian Mounted Police (RCMP) on January 7, 2019, the legal dimension of this ongoing confrontation between the Wet'suwet'en people and the proposed pipeline project, and what this means for the sovereignty of the Wet'suwet'en First Nation and the ongoing work of the Unist'ot'en Healing Center within that territory.

If you recall in my interview with Ramon Elani, we discussed the idea of the home as being a site of resistance. Our rootedness to the land, to what can be properly defined as home in the most sacred sense, informs our passionate resistance to this dominator culture. If we don't even treat our homes as sacred, how then can we even approach treating the earth as sacred? Speaking as a settler, as a white man, I can only say I have a long way to go in reclaiming anything like a home. While in the modern materialist sense, I've certainly never been homeless; I've been incredibly fortunate and privileged in that regard. But to truly be at home is a very foreign concept to me, and for what I can imagine is the case for practically everyone within the confines of the dominant culture of North America.

When I began putting the various threads of this book together, this interview with Dr. Tait really pulled on this notion of not just making a home, but defending those places that serve as a home already, and that have been largely untouched by development and industry in the service of capital. Standing in solidarity with those that hold this space, not allowing the violations of industry to continue, is crucial in these times. The Wet'suwet'en people are resurgent. Their fierce resistance is a potent demonstration what a sacred commitment to the land looks like.

PATRICK FARNSWORTH: I think people are seeing these images or the news and not fully grasping what's happening. They hear about a pipeline and about the Canadian police coming in, but I think there's a lot of details missing. So could you talk about what exactly happened that day?

MOMENTS LIKE THESE, INTENSE MOMENTS WHERE LAND IS AT THE CENTER OF THE ISSUE, SHOW US THAT WE HAVEN'T MADE THE PROGRESS THAT WE NEED TO MAKE.

DR. KARLA TAIT: I think the reason it's getting so much attention and such a reaction from the international community—as well as here in British Columbia—is that it's a current day example of the perpetuation of colonial violence against Indigenous people in spite of all the commitments Canada's made as a country, such as adopting the United Nations Declaration on the Rights of Indigenous People. There's a number of ways that we get partial commitments, lip service about reconciliation with the government of Canada. Moments like these, intense moments where land is at the center of the issue, show us that we haven't made the progress that we need to make.

I'll circle back to the specific circumstances of this confrontation by acknowledging that in the Wet'suwe'ten First Nation, there are five clans and thirteen house groups. We have a complex yet elegant governance system that predates Canada, a hereditary governance system that has been maintaining our traditional territory since time immemorial. The recognized authorities over our territories are our hereditary chiefs.

So we have our Hereditary Chief Knedebease, who has been the Head Chief of our Unist'ot'en house group. We belong to the Dark House, which is Yex T'sa wil_k'us. And that's been since time immemorial. Our law is very explicit about what that authority extends to: the right to control access to our territories, the right to maintain and ensure sustainability of those territories for members. There's also some reciprocal and social responsibility built into our laws: we can't make decisions on our individual territories that are going to negatively impact our neighbors. We saw this with historical fish harvests, in which we had to control how we were drawing on that resource in a way that allowed for our neighbors downstream to still get what they need.

I want to really drive home the point that we've had this long standing governance system that predates Canada. Fast forward to 1997, our hereditary leaders, along with Gitxsan leaders, which are another nation that our territory neighbors or comes alongside, brought to the Supreme Court of Canada this question about land introspection, or actually Aboriginal rights and title. The Canadian Supreme Court acknowledged with their ruling that First Nations people in British Columbia had never ceded or surrendered any of our traditional territories, and they recognized hereditary leaders in their oral traditions and brought forward that fact, how they'd governed and maintained and used the lands.

That decision didn't fully answer this question about title for the Canadian courts. In our minds it's always been crystal clear who has jurisdiction and authority over the land, but for the Canadian legal system, they decided to only answer this partially, acknowledging that we never ceded but not firmly recognizing the title over the land that that chiefs have. So for the Canadian system, we need to bring that question to finality and challenge it in court. Mind you, it took ten years to get to that decision and it came at immense time and financial expense in a system that is set up for Canada's benefit. But we complied to get that landmark ruling, which really paved the way for First Nations to be consulted on industrial projects on traditional territory.

I'm going to skip forward again in time to ten years ago, when my Aunt Freda decided to move back to our Yinta, our traditional territory, and she moved to Wedzin Kwah, where we've established a control or access point, the gate to our traditional territory, in the interest of protecting it from development and the Enbridge pipeline at the time. I think ultimately we were successful in maintaining our ground, because there was a lot of pressure about that type of fossil fuel and backlash from community about sustainability and whether it was even a good investment. There was a few different ways the project fell apart, but that was one of the impetuses for Freda moving on to our territory, and it all really provided us an opportunity as a house group to consider what we actually want this remaining almost 10 percent of our traditional territory to be used for.

That's where the notion of the healing center came up. We had this vision that we could build upon all the support Freda had garnered, and we could actually build the structures and the buildings that would support people to get back out on the land and decolonize, to revitalize culture and build up the resiliency. We know all of that is critical for First Nations people to withstand and overcome some of these challenges we see over-represented in our people as a result of colonial legacies, like alcoholism and intergenerational trauma.

So that brings us to now, and CGL, which has provincial permits supporting its construction and huge government investment from Canada. And also a complete lack of proper consultation and consent from the true authorities on the territory that the pipelines propose to pass through.

PATRICK FARNSWORTH: This is what I suspect is tripping people up. It doesn't really surprise me and I'm sure it definitely doesn't surprise you, but for many it probably seems surprising that this could happen. People are seeing it play out in real time and they're shocked. They're seeing that these people have rights they've proven in court and they have this territory, but having trouble imagining what it would be like if someone invaded your own private property...

DR. KARLA TAIT: Your home, really.

PATRICK FARNSWORTH: Yes. It just sort of surprises people. Justin Trudeau and Canada itself have branded in a very expert way as a very progressive country, a progressive nation where everybody is included, and it's very diverse and where they love their Indigenous people. But it's very obvious that there are certain interests that need to be served in order for the Canadian economy to continue functioning in the way it has. I think a lot of people don't realize how much Canada's economy relies on oil production, and how that violates a lot of Indigenous peoples' sovereignty.

So the story is not just about fossil fuels. While that's absolutely a part of it, it's about so much more than that. I think that's why the story hits a nerve in so many people. You talked a bit about addressing the legacy of intergenerational trauma of colonialism and trying to bring people in that territory back to the land and healing through that. Could you elaborate on that a bit more?

DR. KARLA TAIT: Well, for Wet'suwet'en people—and I hate to pan-Indigenize—but also for many Indigenous people, we center our identity and our culture on the land and our relationship with the land. We've been deliberately removed from and disconnected from our land and culture, time and time again, since colonization. That disconnection from our land leads to many economic and social challenges and struggles for our communities. I think about all of the customs and the ceremony and the practices that we have in our community that make us Wet'suwet'en, and how many of these are linked back to our land in our territory.

I'll share an example from my personal life. My daughter is quite young, and since Freda has moved back to our territory it's really prompted us to do a revitalization of our cultural practices within our Unist'ot'en house. This has meant that we are coming back to engage in our annual action camps which focus on reconnecting and protecting our traditional territories, and meeting with other Indigenous people around that, as well as other aspects of wellness and strengthening that connection.

We've had this cultural practice, the ceremony for our people, that we've just brought back into practice within my house group through my daughter. I think she's the first one within two generations that we've carried out this really ancient practice with. It's a ceremony that really connected her to the land after her birth. I won't share the finer details of it, but it really exemplifies the fact that we see our members as one and the same as our land.

So this was a practice to connect her and root her in our territories. I've seen its impact on her own identity formation, how it's potentially going to build up her resiliency in a way that I didn't have the opportunity to have. She's at a really young age, and she is both Lakota and Wet'suwet'en, so I'm mindful of some of the teachings of both of those cultures. From the Lakota teachings, she's going to be a spirit until she's six, and will be very connected to the spiritual realm up until then and it's our responsibility as parents, family, and the larger family system to take care of her, to nurture her so she decides to stay on this Earth World and doesn't decide to return to the spirit world.

WE'VE BEEN DELIBERATELY REMOVED FROM AND DISCONNECTED FROM OUR LAND AND CULTURE, TIME AND TIME AGAIN, SINCE COLONIZATION.

I love the alignment with some of the teachings in psychology regarding that critical developmental period, up until six. We know that from psychology how important that period is, and it's just fascinating to me that ancient Indigenous knowledge and knowing were already there prior to the scientific backing of our Western theologies.

I think it's just so clear that she is connected in that way, and I don't know if that's part of what this ceremony did, the intrinsic knowing of that ceremony that literally helps connect a part of her to the land. I saw it from our early visits to our territory, in her babbling and her attempts to engage us as adults in our lives. That developmental stage is where we expect she's going to engage people in the room, and then try to build that connection, that bond, that attachment. I saw those kinds of behaviors as we were walking through the territory. She'd increase that babbling: she looked like she was speaking to ancestors on the land and it was only at the sites within our territory where there historically were houses or gathering places that our community used.

I think about how historically important that connection to the land for our people was, and how that must have shaped us and made us strong. You know: to know that we aren't ever alone in this world, that we have land to ground us, that we have ancestors behind us, and how much that prepares an individual to face some of the overwhelming hardships. Especially for Indigenous people, all of the ways that we experience inequities in social determinants of health and poverty.

It's a big issue, and I don't know if I'm giving it justice, but we're keenly aware of how important that connection to our land and our territory is, as well as being able to grow up experiencing our culture in a really rich, direct way: trapping on the land, fishing off the land, drinking pristine water from a river right on your territory. It helps keep us grounded and connected. When individuals and your nation are still reeling from historic and current experiences with colonization and disempowerment, how important it is to be grounded in something much bigger than ourselves—the land!

Freda summarizes it really nicely in some of her quotes, acknowledging that when we heal the land, we heal our people. There's not a dichotomy there. We're one and the same.

In spite of the overwhelming, blunt force wielded by the colonial state of Canada and its corporate partners to push through this pipeline project, the Wet'suwet'en and their allies have defended their territory. Their outright refusal to allow their lands to be violated by industry is a testament to their strength and resolve. As Dr. Tait expressed so beautifully, too much is at stake to even consider submitting to the demands of the fossil fuel industry and the State in service to capital.

But just as Indigenous people continue their 500-year-long anticolonial struggle for sovereignty, which is really synonymous with protecting life itself, the other side of this sacred response is to regenerate the land that has been desecrated by industrial civilization. So much of our planet has been violated in service to the logic of capital, whether directly through relentless extraction, or through the collateral damage incurred by pollution, erosion, drought, destabilizing climate shifts, ecological collapse, and mass extinction.

We, the inheritors of this culture, have a great task ahead of us. We need to reclaim the ways of being that our ancestors abandoned on the path to modernity, despite not knowing whether we fully succeed. I am reminding myself while writing this, and have been reminding myself while pulling these threads together: we likely may fail. Nevertheless, what the hell else are we going to do with this time we have? Our sacred obligations remain, in spite of all that has transpired—and what will come—in these times of trouble.

JOE BREWER

The dominant culture robs the planet of its regenerative capacities. What would a culture look like that would enhance the Earth's regenerative capacities, instead? There are numerous individuals and organizations that are attempting to answer that question right now, doing what I've called invisible, sacred work. One of the leaders in this work is Joe Brewer, who has described himself as a change strategist, complexity researcher, cognitive scientist, and evangelist for the field of cultural design.

When we recorded this interview, Joe and his family had relocated to Costa Rica from the United States to engage directly in the implementation of "regenerative hubs" networked across the planet's numerous bioregions. These hubs serve numerous purposes, including providing the scaffolding of knowledge and practice to manage planetary collapse as nation-states, economies, ecologies, and the global climate system continue to break down.

I ask Joe about the necessity of doing this work in spite of the overwhelming likelihood that human and nonhuman life will go extinct as a result of ecological collapse and abrupt climate disruption. Even in the face of planetary collapse, there are individuals that are doing (seemingly) invisible and sacred work in managing this unprecedented collapse, generating and expanding systems that heal the land, the climate, the human psyche, and humanity's relationship with the living systems of this planet. As Joe says in this discussion, what can we do that is worthy of our hope?

PATRICK FARNSWORTH: I feel like there's two contradictory things here. By managing planetary collapse, there has to be a collapse for the types of systems you're talking about to exist and flourish. But in that collapse, it becomes increasingly difficult to promote these types of ideas, because censorship increases. It becomes difficult for people to have access to this information, and more and more people will be caught up in the immediate crises. There's going to be all of these stresses that will be put on people.

I imagine that, from your position, the question is how to build programs and systems that are regenerative and also very robust even in the face of planetary collapse. And hopefully

THE DEFINITION OF BEING ALIVE IS A DYNAMIC PROCESS, TO KEEP REGENERATING THE CONDITIONS FOR IT TO CONTINUE LIVING.

people will still be able to find you and others that are working on this. Hopefully they can find these programs, these systems, these ways of understanding and being in the world, and be able to act on it within the context of a planetary collapse. That seems like the most difficult thing to grapple with.

JOE BREWER: Yeah, it is something I grapple with on a daily basis. One of the important things to know is that, if you start to embody regenerative practices, you're actually becoming more healthy. That's because regeneration is a feature of living things, and we're living things. The definition of being alive is a dynamic process, to keep regenerating the conditions for it to continue living.

The moment-to-moment existence of life is the phenomenon of regeneration. When we understand how deeply that is true, when we understand that depth, we start to see when to we do regenerative things, our health improves, our stress levels go down, our ability to discern increases, and our ability to take effective actions goes up.

Right now, one of our big problems is that we have all these debt-based currencies and a global financial system based on servicing debt. What if more of us definancialize our lives, stop paying the debts, walk away from it, and stop trying to earn money because all money is debt in the current system? Instead, we could set up and use alternative currencies when the current system becomes dysfunctional. This is the real beauty of blockchain, hollowchain, digitally autonomous organizations, and the whole conversation around token economies.

It's a big conversation happening right now: different ways of accounting for, measuring, and validating value within a system. These things will let us walk away from the disaster of the current financial system and do something else instead. People are feeling the strain of the pathological economy. It's all about extraction, that's why it's pathological. Some are instead setting up value exchanges within their local web of relationships, tracking those value exchanges in other ways than fiat currency: using an alternative currency. You can do a time banking and say, "I give an hour of time, you give an hour of time." This can be done with bartering, with actual objects being the things of value. As long as they are not using debt-based money, they're moving themselves closer to resilience.

Most of us don't realize we can choose not to use debt-based money. We can just choose at any time and declare it invalid. One of the reasons we don't do this is because we don't know how to take concrete actions if we don't have money. Most of us have to have money, because the only way we get food is to buy it at a store and they only accept money. So until you can work out an agreement with the grocer to give you produce or food without giving them money, then you feel you must have money. The need for it feels real, but it's a forced artificial environment that makes it feel real.

Where I'm living now, there is livestock, cattle, pigs and horses, and chickens. There's also a lot of organic gardening and regenerative agriculture that's happening. So where I'm living, I don't have to pay for the food, because the food is just part of the production system of where I live. If I'm doing something valuable for the community here, people who are working in the food production part of it are benefiting from whatever I bring, and so they're happy to just keep producing food and I get to eat some of it. And so I don't have to pay them any money for that. So, if we can build these closed loops of value creation, we create value and different relationships, and they form loops with each other where nothing is being extracted, it's always being recirculated. This is one of the principles of regenerative economics: robust circulation. Robust circulation is like the circulation of blood in your body. If value creation in the economy is circulating like the blood in your body does, then every piece of cell tissue in your body gets nutrition. In the economy, every person in the economy gets their needs met.

The more we engage in regenerative practices, the healthier we become. This is a really important lesson, because as more of us start doing this, our neighbors are going to notice when they have hardships. They'll start coming to us, because we're the ones who can still help them. If you need food and the only person who has food is the farmer doing regenerative farming, then you go and say, "can I get some eggs? I really need some help." And that farmer might say, "well, I actually need some help with the harvest. Could you come help in the field?" And now you're having a meaningful exchange that doesn't involve money.

We tend to fall back to agricultural examples, because they're so important and they're so easy to visualize. But this applies to anything in the economy. We need to build those relationships of trust and value exchange, so that there is robust circulation, which means we have a healthy set of relationships and a healthy community. This needs to be taught, and people need to practice it, experience it, and really learn it.

PATRICK FARNSWORTH: I am becoming increasingly aware of this invisible work happening all around us, under the radar. People are doing and working on things, they have ideas and systems they're developing for when things really do fall apart. While it may not be fully functioning yet, people know what's happening, and they feel a calling to do something about it, and they are in service to something bigger than themselves.

I would describe these as invisible acts of love that hold everything together under the surface. Your work is invisible in many ways. Many people aren't really aware of what you and others are doing. I like to highlight that, because as these systems begin to crumble, the work you and others are engaged in is going to emerge. It's going to be apparent to people that, when they can't go to the grocery store anymore to buy food or when their fiat currency isn't a viable means of transaction any longer, what you and others are building become visible.

I think this invisible network of regenerative and collapse-aware people is a really remarkable thing.

JOE BREWER: Yeah. It's really remarkable because of the way that the dominant discourse has been framed. I'll use an example: Greta Thunberg. Right now, she is really getting a lot of attention. She is eloquently speaking about the climate crisis, but she has some very problematic frames. One of these problematic frames is that the only mechanism we can address the climate crisis is through the decisions of elected politicians. She keeps saying, "please, people who are elected, make the right decision," and "please, people who are voting in the next election, elect people who will make the right decision."

That framing of elections as leading to the needed outcome is so disconnected from all of the history in the last ten or fifteen years, the history of propaganda, social media, surveillance capitalism, captured systems, and clandestine tax evasion networks. It is so blind to all the structures of power, and it's basically giving up our personal sovereignty and saying: "I'm not a sovereign person, but at least that elected person is. So maybe I can beg them and they will do the right thing." That's very different from doing regenerative work and raising a

family around regenerative practices. That is being sovereign: making decisions for yourself.

That's one framing problem in the climate discourse that Greta is embodying. Another one is the idea that what we need to do is deal with CO_2, as if carbon dioxide is the primary problem, as if it's going to cause humans to go extinct first. Most people don't know this. If you ask a roomful of environmentally-aware people what will likely to cause humans to go extinct first, most people would say climate change. But that's not it. The thing that would cause us to go extinct first is the loss of biodiversity and the unraveling of ecosystems caused by land degradation.

Obviously, climate change contributes to the loss of biodiversity, but it's actually the destabilization of the biosphere that's the primary problem. One subtle way to think about this: a lot of the biodiversity is in the microbes in the soil. When we think of biodiversity, we might think of big birds, cool looking insects, and other things that come to mind because of the way the human cognitive system is biased. But the majority of history of life on Earth was protozoic. The majority of history of life on Earth was bacteria.

The majority of diversity of life on Earth is bacteria, but we have not really been able to measure it because it's so complex to measure. So as we pour all of these poisons, pesticides, and herbicides onto the soil, we're killing the biodiversity of soils. That is going to cause our death through the collapse of food systems: it will cause us to starve. That can cause humans to go extinct much more quickly than climate change. We've framed climate change as separate from biodiversity, which is its own other level of problematic framing. But the idea that CO_2 emissions are what's going to cause human extinction or the collapse of our civilization is so misguided: it's a symptom rather than a root cause.

How could anyone get to regeneration if they're looking at CO_2? CO_2 is a technocratic term, a chemist's term. I have a master's degree in atmospheric science, so I understand the importance CO_2. But the framing of this whole thing as a CO_2 level is the problem. The problem is that humans' social niches destroy the conditions for life. We need human social niches to be regenerative, to be creating the conditions of life. And until people can get that, they won't be able to see how there could be a solution, and this movement will remain invisible.

PATRICK FARNSWORTH: Cory Morningstar's research into the nonprofit industrial complex was to address how propaganda and social engineering are being used to reframe the discussion into how governments and corporations can save us. That is not the way we should be framing the discussion at all, and that's why I say this invisible work is so valuable. There are real people doing real things right now, and that isn't invisible at all. It's just that we live in a paradigm that blinds us to it.

JOE BREWER: Let's take that rabbit hole a little further and notice that the power of our blinders goes in both directions. Those of us doing this radical subversive work are going to go unnoticed by the powers that be, because they don't get it. Anyone who is in a life-destroying, psychopathic, extractive mindset is not going to understand the power of regenerative community, and they're not going to see it sweeping things if it succeeds in getting a foothold. For example, when the country of Ecuador declared that nature has equal rights with other rights-holders in its constitution, it was like spiritual jujitsu. We just change reality with that statement, and the power of doing this is so intense.

We are sleepwalkers dreaming a nightmare into existence, having no idea that we're the gods creating it. So we think that all these horrible powers, these huge corporations, all these terrible things are doing it; but it's humans doing this. We're humans. You're a human. I'm a human. We have this incredible cultural power, and if we learn how to use it, if we learn how to be mindful of it and to work with it, then it's a totally different situation.

WE ARE SLEEPWALKERS DREAMING A NIGHTMARE INTO EXISTENCE, HAVING NO IDEA THAT WE'RE THE GODS CREATING IT.

The earth has regulated itself, it's had homeostasis for 3.8 billion years of life, until humans have come along. There's only one other time where something really screwed up the balance: oxygen-producing bacteria, which paved the way for the eukaryotes. But since that time, there has never been an organism on Earth that was capable of changing the self-regulation of the entire planet.

Here we are, we're the first to do it in this profound way, using culture to do it. When those bacteria were releasing oxygen, they were creating their own demise unconsciously, because they did not have a cognitive system that could let them see it. Humans have cognitive systems to let us see it. We created a global network of earth-observing satellites. We are watching the Earth change. We built the systems of perception for the planet.

So this means that we as human beings can become the immune system of the planet. We can become the homeostasis of the planet. And that's not a metaphor. It's a literal description. We can be the functioning management apparatus on a healthy planet. That's how strong our cultural abilities are. And our blindness to this is so severe because we can only see the bad that we're doing, and we don't see the beautiful power of what we are.

PATRICK FARNSWORTH: Yes, absolutely. I feel like the climate crisis could potentially be the very thing that kills us, or it could be an initiation into that role. We've developed the tools and the cultural capabilities to do this, but it's just a matter of getting there.

JOE BREWER: Right now we need really powerful stories. We need stories that tell people: quit your job and do this instead. Move your family to a different place, if that's what you need to do. Don't go to college. Go learn in one of these regenerative campuses instead. Don't accept the rules of the game, play a different game, and have that game be based on reality.

Because this is our shot. We've got about 15 or 20 year time window, and the first thing we need do is stop losing biodiversity. Second: we need to start regenerating degraded land. Those two things are the same thing, by the way. And as we regenerate degraded land, we store all of that extra carbon in soil and in the embodied carbon of organisms in the ecosystems. We convert it back into biomass. If we do that, we are regenerating the earth, regenerating ecosystems and restabilizing the planet.

Anyone who says it's about having cleaner pipes on our coal power plants, or replacing one kind of single occupancy vehicle with an internal combustion engine with another one that runs on electricity—which is actually powered by a coal power plant—doesn't get it. They just don't get it. And don't waste your time trying to get them to get it. We don't have that much time.

Joe is engaging in invisible, sacred work, laying the groundwork for regional hubs that enhance the co-creative capacities of human beings in relation with the life-generating capacities of the land. This work is crucial in developing bases of ecological knowledge and cultural frameworks deeply needed to manage planetary collapse and build a viable, practical means to do so.

When Joe discusses managing planetary collapse, it points to a few things. Industrial civilization is meeting its end, and operating within the logic of the system cannot, and will not, generate the level of radical response and organization needed to sustain human life in any form into the near and distant future. Our collective fate is intertwined with the regenerative capacities of the living planet, and only by using our creative gifts as a species will we have anything like a chance of navigating our way through this.

JOHN HALSTEAD

To face what comes requires humility. We will not win. Our civilization is dying, and maybe it should. As with all things, we exist within grand cycles of life and death, and we have numerous mythologies and spiritual traditions that point to this. It is with this in mind, I present pagan writer and former activist John Halstead, in an interview I conducted with him after reading his two-part essay published at Gods & Radicals, "What If It's Already Too Late?: Being an Activist in the Anthropocene" and "Die Early and Often: Being Attis in the Anthropocene."

In particular, what I find intriguing and especially useful is how John applies a pagan perspective to the subject of the human-caused ecological crisis unfolding presently, guiding the reader to a mythological understanding of humanity's role within the grand cycles of life and death. In his essay, he uses archetypes presented in the myth of the dying god as a framework to process civilization collapse and the destructive role humanity has played within the living systems of the planet. In acknowledging where we stand within this process, we can begin asking the right questions. To quote from "Die Early and Often: Being Attis in the Anthropocene":

> This means more than merely accepting our fate. It means shifting from an egocentric perspective, in which death is the ultimate evil, to a cosmic perspective, in which death is part of the cycle of life. Rather than raging against the dying of the light, we surrender to the Wheel, and in surrendering, we are transformed. This transformation brings no apotheosis, no individual immortality, but it enables us to realize the meaning of our lives as part of a greater whole which transcends us.

PATRICK FARNSWORTH: In the essay you mention you went to a climate rally and there was the founder of 350.org, Bill McKibben. Bill McKibben has been at the forefront of climate activism. He's been definitely the one that's been speaking the most loudly and has been organizing the most. I read in the essay that, when you heard him speak at this rally, that was really when you understood we're in a phase in this climate change scenario where it's already too late? What was that like for you?

IN GENERAL TERMS, PAGANISM IS A RELIGION THAT LOOKS BACK TO PRE-CHRISTIAN RELIGION AND MYTHS. IT DRAWS ON THOSE MYTHS AND THOSE RITUALS IN A CREATIVE WAY TO DEAL WITH MODERN PROBLEMS.

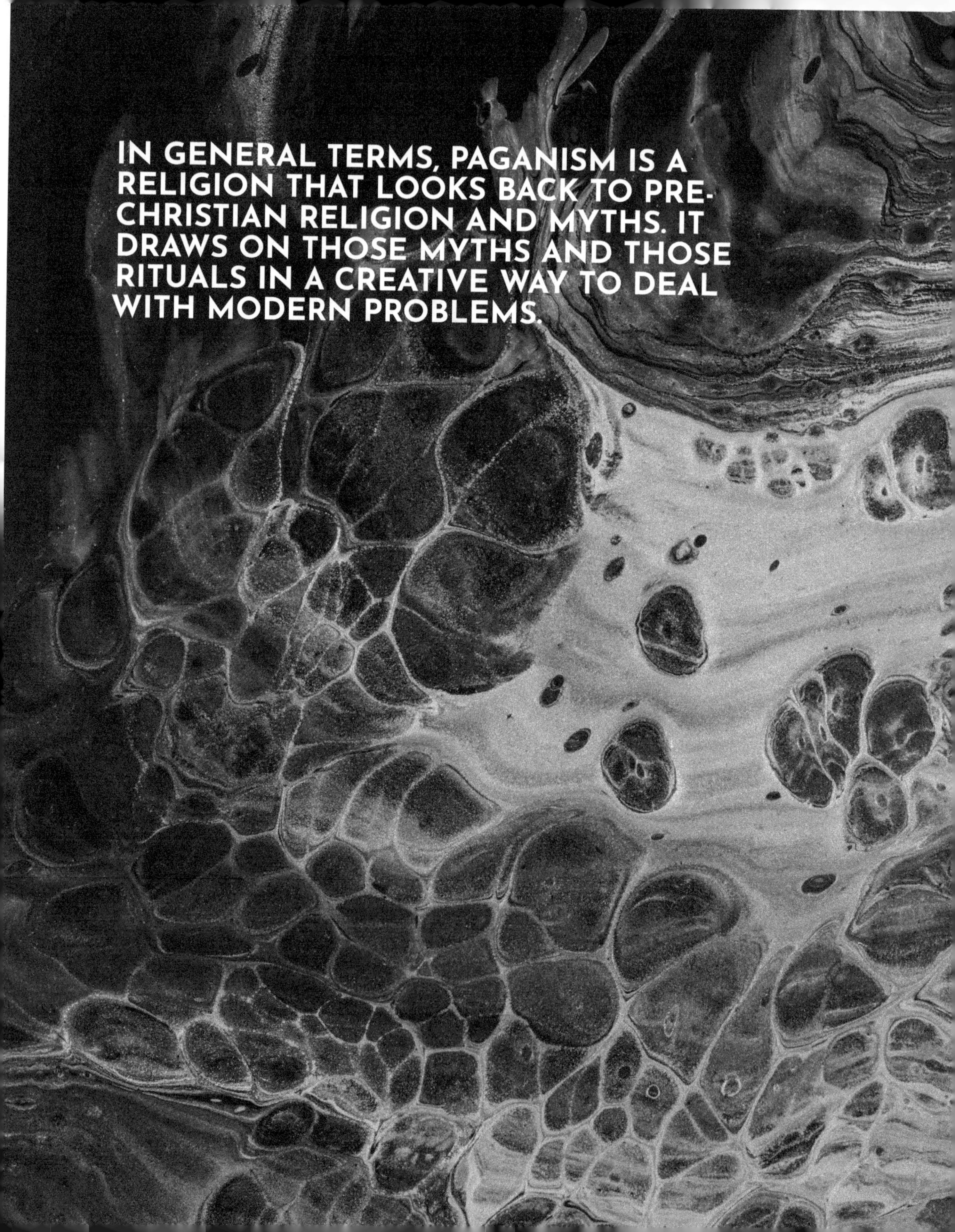

JOHN HALSTEAD: So it's probably one of the biggest rallies I had been to. Bill McKibben is there. It was a big deal and I was excited to hear him speak and what he said stuck in my mind. I had to find the recording to make sure I had heard him right, because it's been a couple of years. What he says is we're not going to stop global climate change. It's too late for that. And to be honest, at the time I was getting ready to get arrested and I was real excited and pumped up.

I put it on the back burner: it went in that part of your brain where you know it happened, but you don't want to think about it. And it stayed there until very, very recently. It would pop up in my mind occasionally and nag at me, and I don't think that's unusual. I think that happens to many environmental activists probably, in the dark of night when they're alone and not chanting, "when we fight, we can win." Probably most environmental activists ask themselves that question at least once, or probably frequently: "is it too late?"

It was actually a book that opened the door and let me seriously consider revisiting what McKibben had said. The book was *The Wizard and the Prophet*, by Charles Man. It's an interesting book—a story about two different scientists and their competing visions of our relationship to the earth, specifically regarding food production. One of them was Norman Borlaug, who has been credited with the Green Revolution. Some people say he's saved a billion people from starving. On the other side, some people say because of the industrial farming practices that he inspired, he's killed a billion people. So he's either a savior or a mass murderer. And then on the other side was William Vogt, who took a different track by saying that technology is not going to save us. What we need is to scale back, to conserve and cut back on the human population.

The author uses these two scientist as archetypes. He walks through various areas of our relationship with the world of air, water, earth, and energy, looking at different scientists and putting them in different categories, asking, "are you a Borlaugian or are you a Vogt-ian?" At the beginning and at the end of the book, the author addresses a third perspective, that of Lynne Margulis, who's the microbiologist famous for, with James Lovelock, the Gaia theory. She came up with the theory that some parts of ourselves—like our mitochondria—used to be independent organisms and, through symbiosis, these different organisms came together to become one organism. So, at our root, we have interdependently existing organisms.

Lynne Margulis says we're doomed, just like every other species on the planet is doomed. She studies microorganisms, and she says it's just the nature of life to expand until you hit the edge of the proverbial petri-dish, and then that's it. And so the author doesn't really deal with her perspective because he's bouncing between two questions: will technology save us or do we need to contract and conserve and reduce our population? Which one of these is gonna save us? And then he throws in there Margulis saying: "we're not gonna be saved."

Borlaug is the wizard and Vogt is the prophet, and I found myself on the side of the prophet saying we need to constrict. I don't believe technology is going to save us. But what about this idea, what about what Margulis says, that there is no salvation, that we are going to die? So I started doing some research, pulling things together, and more and more that seemed the perspective that made sense.

I recently went back and remembered that Bill McKibben was saying pretty much the same thing a couple of years ago. I went back in to see if I was remembering him right, and sure enough, I had. It was shocking. Think about the founder of 350.org at a climate rally saying, "it's great that you're all out here, nothing is more important, but we're going to lose." He was not inspiring.

PATRICK FARNSWORTH: [laughing].

JOHN HALSTEAD: It kind of fell on deaf ears that time. At least mine. And it came back later.

PATRICK FARNSWORTH: I think using the language that we're "fighting" climate change, this war language, this battle mentality, makes it as if climate change is an object that we can beat back and tell to "stay away." We want to keep on living, so we'll keep on fighting. But that's not really what I think climate change is asking of our species.

Maybe all these things we're seeing in the news indicate something else that we're all afraid to admit: this civilization is facing very serious threats to its survival and our way of life is going to end. I think the difficulty is getting from point A, which is that climate change is real and it's a problem, to the point where you and I and others are coming to, which is that it's too late. That is a very frightening place to be. Many people don't want to get into that space, because they feel like that's defeat.

For most people, this is a very unusual position to be in, I think. It makes them very uncomfortable to even have this conversation. And for some of those that are willing to have the conversation, they bring an antihuman sentiment: human beings are a rotten species and this ecological crisis proves how rotten we are. You don't directly address that point in your second essay, but you bring up this really fascinating mythological, pagan perspective. I really love this perspective that you provide on this death that we're experiencing, and I would love if you could discuss the mythological perspective more.

JOHN HALSTEAD: Thanks. I was looking at the statistics for the two articles, and they're supposed to be a two-part essay. But about three times as many people read the first article—which is about how we're doomed—as have read the second article—which is about, what do we do now? That's a little disturbing to me.

The most attention I've gotten, the most responses I've gotten have been about the first article. It was really a setup for the second article, which is, what symbolic resources can paganism offer to help us respond to the question of "what now?" You actually set it up pretty well when you talked about the battle mentality of environmentalism. I think I see that same thing, too, talking about "fighting" climate change. It's not that different from the industrialists whom we perceive as fighting nature, fighting against the limits of nature. There's this idea that we can struggle with nature and win, and this idea is at the heart of what got us into this place. It's also at the heart of a lot of environmental activism.

I REALIZED I WAS GOING THROUGH THE STAGES OF GRIEF WITH MY ACTIVISM. YOU KNOW, THERE'S ANGER, DENIAL, AND BARGAINING, AND I CAN SEE ALL THIS IN THE VARIOUS THINGS THAT I HAD DONE.

I realized I was going through the stages of grief with my activism. You know, there's anger, denial, and bargaining, and I can see all this in the various things that I had done. In a way, it's not that different from climate change deniers. In a way, both sides are denying what climate change really means for us as a species.

So to get to get your question, I grew up religious, I was raised Mormon. Soon after I left the Mormon Church in my twenties, I found paganism. I've always been drawn to religion, and I've always been a little frustrated that the environmental movement was so secular. There's almost an antireligious attitude: even people who are personally or privately religious don't like to bring it into environmentalism, because they're afraid of either alienating people or being alienated. From my perspective, I don't believe that human beings are really rational beings. If we were, I think we would have, back in 1988, started taking serious action to reverse our course as far as carbon emissions.

I don't think people really change their behaviors or come to a radical change in their life by facing facts and figures. I think it's more right-brained than that. I consider myself to be a very analytical person, and yet I often find in my own decisions that, though I think I'm being rational, they are actually post hoc rationalizations. I really made the decision based on something more holistic, something more right brain. So for all its problems and for all its limitations, I think when religion is doing what it does best, religion is really good at dealing with right brain stuff. It really speaks to that part of our brain that responds to story or myth, to symbols and to rituals. When the environmental movement and environmental activists cut themselves off from religion, they're cutting themselves off from a lot of resources that can be drawn on in the climate movement. Now, occasionally you'll see Indigenous peoples being invited to invoke the sacred at the beginning of an environmental action. And I've always thought that was great, but I'd like to see some more of it.

Like I said earlier, I was a little frustrated with my own paganism, because as I was finding that any religion has its good points and bad points, its own limitations. And I was frustrated that I didn't see the kind of environmental activism among pagans that I expected to see. But I still nevertheless think there are a lot of resources that can be drawn on, symbolic resources, from a variety of religions including paganism.

In general terms, paganism is a religion that looks back to pre-Christian religion and myths. It draws on those myths and those rituals in a creative way to deal with modern problems. There's as much variety among pagans as there is among Christians, so you can't really generalize very much, but depending on the type of paganism you're talking about, there's going to be creativity in drawing on these ancient sources.

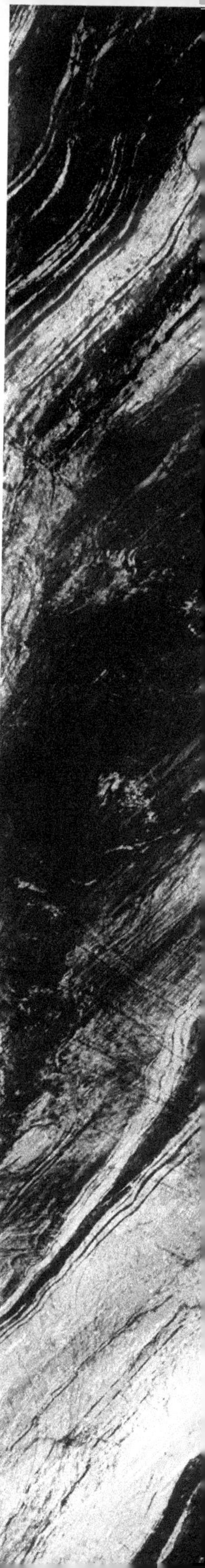

One of the symbolic concepts that drew me to paganism was the myth of the dying god, and this is what I talk about in the second article. The dying god myth, actually though there are ancient examples of it, this archetype was really an invention of James Frazer, who wrote *The Golden Bough*. Frazer's unstated goal in writing about the dying god was to debunk Christianity. He wanted to show all these ancient examples of gods that died, and therefore wanted to imply that Christianity was bunk because the Christ figure is just another one of these ancient dying gods.

The basic idea in its modern form is that there's a god who is born to the gods on the Winter Solstice. Over the course of one year he matures, mates with the goddess, and then is killed or dies and then is later either reborn or reborn in the form of his offspring. In the winter solstice he is born, in the springtime he grows up, in the summer he makes love to the goddess, in the fall he is sacrificed, and then he is reborn in the winter.

What does that have to do with really anything? The underlying message of that story is that death is a part of life. It is a necessary part of life, and without death you don't get life. This is very literally true. If you've ever composted, you experience this in a very visceral way: life feeds on death.

This myth is very different from the myth driving our culture. Our culture is built on this myth of progress. There's no death, we are continually triumphing, battling back the forces of nature.

This myth presents a very different paradigm for our society. When thinking about what we can do in the face of climate change, in the face of the fact we very well may not survive as a species, it occurred to me that this myth of the dying god offers some resources. What the myth of the dying god tells me is, "of course we're going to die. We were always going to die." Even if you want to project it out as far as possible, even if humanity survived a millennia or eons, eventually our sun will burn out.

Every other species has at some point passed away, and it is tragic. That's a part of the myth of the dying god. It doesn't diminish the tragic nature of the sacrifice, the tragedy of death. But it's unavoidable. And we can have one of two responses to this reality. One is to embrace it, and one is to deny it. Our society, by following this myth of progress, has been on a track of denial. We deny our own mortality, and in myriad ways we deny our collective mortality. You can think of human civilization as a giant exercise in denying our mortality.

I tell the story in the article about my son. When he was thirteen he was starting to question Mormonism. Among other things, he was really distressed about the possibility that he was going to die. You know, I tried to help him as much as best as I could, because I don't believe in an afterlife and I tried to help him come to terms with it. Where he came to rest is where I and a lot of atheists and religious naturalists come to rest: we're going to die, but at least our civilization will continue to carry on and the accumulated knowledge of all humanity will just keep rising. And that's not true either. Turns out that's a myth. Every other civilization before us has died and ours is going to die too.

The real tragedy is that our civilization is going to die in a spectacularly destructive way. We're not just going to take out some local bioregion or the Nile Delta. We're going to take a good part of the planet. Life is going to survive, but we're going to take out a good part of biological diversity on our way out the door. And that's a whole new level of tragedy, but accepting it and then figuring out a way to live meaningfully in that acceptance is the challenge that we have. I think for me, the myth of the dying god helps me hold that truth, to look at it, and then consider how to live meaningfully in that context.

THE UNDERLYING MESSAGE OF THAT STORY IS THAT DEATH IS A PART OF LIFE.

DAHR JAMAIL AND ROB SEIMETZ

At the very beginning of this book, I featured an interview I conducted with journalist and author Dahr Jamail, in which he lays out the data surrounding abrupt climate disruption and where this trend will likely lead. Since that interview was conducted, Dahr's book *The End of Ice: Bearing Witness and Finding Meaning in the Path of Climate Disruption* was released, in which he skillfully and beautifully elaborates the dire consequences of human-caused climate disruption, including on-the-ground research into the rapid and accelerating impacts of climate disruption in some of the most dramatically affected places on Earth. These places include one of Dahr's favorite peaks to climb, Denali in Alaska; Camp 41 in the Amazon rainforest under the stewardship of world-renowned ecologist Dr. Thomas Lovejoy; and the Great Barrier Reef off the coast of Australia, undergoing a massive die-off as a result of warming oceans and ocean acidification.

Dahr met a diverse set of scientists and researchers all over the world engaged in their own direct research into the impacts of climate disruption and expresses the deeply felt sorrow, anger, and grief that he, and many of the scientists he interviewed for this book, are feeling in the face of the mass extinction currently underway.

My friend Rob Seimetz and I (who was also featured previously in this book in our discussion with Michael Sliwa) collaborated on an interview with Dahr after meeting him for his book release event at Powell's City of Books in Portland, Oregon in January 2019. I want to feature another interview with Dahr because understanding the data of climate disruption and ecological collapse is only the beginning of really getting at what it means to be alive in this time. It's crucial, I think, to discuss the enormous difficulty of responding to this information, both on a personal level and on a wider collective level, particularly within the context of an oligarchic corporate capitalist system that puts the continuation of "business as usual" above everything else.

Dahr, Rob, and I discuss how we have obligations as a species to do the right thing, no matter how tragic and difficult this is in face of what's unfolding on this planet. Doing the right thing doesn't mean that we believe in an outcome that's going to be beneficial for us, that is going to save us. Doing the right thing, doing the sacred work, connecting to the land, connecting to all of the qualities of the material world of which we are inextricably a part includes not looking away from what this civilization has produced.

Learning to say goodbye is a big aspect of this sacred duty. More than anything, this discussion with Rob and Dahr is about acceptance: acceptance of the predicament we find ourselves in, as well as the tragic and sacred duty we have to learn to say good-bye to these places and beings that are disappearing. Within that space of acceptance, we can proceed into our uncertain future with maturity, purpose, and clarity.

PATRICK FARNSWORTH: There's this completely deeper, much more difficult question to ask ourselves: how did we build a society and a culture that was so messed up in so many ways? It's not just with our environment, I think it's a reflection of what's happening internally. I was just having a discussion about intergenerational trauma, thinking about all the various ways our culture has been built on faulty premises. Maybe there are values that can still come out of it, but it seems to me that this crisis is asking something that is so deep and so profound, and I don't know if people are going to come to that realization.

I FEEL LIKE THIS IS AN AGE WHERE WE'RE GOING TO HAVE TO LEARN TO SAY GOODBYE

We must figure how to live beautifully in the time that we have. It's going to be really hard to do that when we have these ridiculous political institutions and the corporate capitalist mindset that runs the economy. It's so overwhelming, and this podcast has been almost a blessing and a curse...

DAHR JAMAIL: I understand—I feel exactly the same about my climate writing. What you just articulated is this simultaneous world-living that all of us have to engage in now. Here we are—we're dry, we have our needs met right now, but also have the simultaneous knowledge that people living on coasts are having to leave because they're flooding. Climate change is upon them. Some live in places where they've already lost everything, already living in the climate-disrupted future, while the rest of us haven't yet experienced a direct extreme weather event that's completely altered our lives. We haven't gotten there yet.

For us, there's still this sense that it's not here yet, it's in the future, it's out there. Yet for increasing numbers of people around the world, it's here. They're living in what we perceive to be the future. At the same time, those of us who have watched this coming and really understand it deeply are getting a chance to emotionally process, like we're on borrowed time. We're getting to let this sink in before these extreme weather events hit us.

I WANT TO LIVE MY LIFE IN A WAY THAT SHOWS MY PRIORITY IS LIVING IN RIGHT RELATION TO THE EARTH.

In the Pacific Northwest it is already starting to happen, with the smoke from the wildfires in the summer and the acidifying oceans hitting the shellfish industries. It is already starting to happen, and the question is, "how do you live in all these worlds simultaneously?" I will leave this interview and then get in my car and drive back home. This is quite the psychological tight rope all of us have to live in. For me, I have to spiritually evolve and constantly be talking to people about what's going on, processing grief, sharing it, and laughing and crying and understanding. This is a truly mad time to be on the planet. I can't imagine a more intense time than right now.

ROB SEIMETZ: I feel like this is an age where we're going to have to learn to say goodbye, and we are going to have to get out in nature to say goodbye properly. You were interviewed on Counterpunch Radio a couple months ago, and I remember that you were wondering aloud whether there are people making calculated decisions on where to move based off of what's happening with climate change. I'm one of those people who did.

I lived in Florida, and after Hurricane Irma came, my wife and I looked at each other and said, "never again." We moved out here, and I was almost in denial that this was a global situation. You see pictures of Oregon and it's beautiful, but the silence was deafening. I haven't connected with wildlife here yet like I did in Florida. But you have to find the silver linings: you talked about last night how you were diving the Great Barrier Reef, and there are areas where it's bleached and areas where it's not. The one silver lining I have found here so far is a hill called Memaloose Hill. It's about 60 miles from here, and in the spring there are flowers everywhere, and you see bees everywhere pollinating the flowers. It's just beautiful. If we're aware of these things, we're going to be able to say goodbye properly. So do you feel this book chronicled those moments for you?

DAHR JAMAIL: I do. I feel like the book provided me with this extraordinary privilege. I was able to go to these amazing places in the world, like the Amazon, the Great Barrier Reef, Glacier National Park, Utqiagvik Alaska (formerly known as Barrow): all these really incredible places that are changing before our very eyes and, in a lot of ways, many of them are going away. There is no question that the entirety of South Florida will be underwater before I die. Or look at what's happening already with the melting of Greenland and the Antarctic, and go to these places fully cognizant of all the science about what's happening and what's projected, knowing that no government on the earth is reacting the way it should be in the face of this crisis.

When I was at the Great Barrier Reef, I really felt like I didn't see myself coming back there. I felt I was saying goodbye to the reef, I wrote about that in the book and I feel like that was a tremendous privilege. How many worlds do we need to try to live in simultaneously, appreciating the beauty, being with it because it's still here and we don't know when it will go away? In the case of the Great Barrier Reef it looks like it's single digit years at this point for sure. It certainly looks that way. But it's still here.

When I'm writing a hard article about the climate and I get overwhelmed, I can step outside my house and be surrounded by big douglas firs, hemlocks, and cedars and just be with the trees. I can find solace in that: they're still here, I'm still here. What I've been saying now on tour is, "if you're still here I can still be here." That's also reason enough to keep doing whatever I can, regardless of how bleak it is, how intensely it feels like we're doomed. I still feel morally obliged to do everything I can to help take care of the planet, because we don't know for sure what's going to happen. Even if we did, I would still try to do the right thing simply because I want to go out on my feet, I want to have dignity, and I want to be in right relations with the planet.

ROB SIEMETZ: I've tried to use sports terms in my writing to try to get people to connect. One term I used is that we need to be gunslingers. When you think of a quarterback, he's a gunslinger, he'll take these big chances to win the game. The game we're trying to win right now is a little bit more time with nature. Maybe it will go away, and there's nothing we can really control anymore. But if we were to be gunslingers with the environment in mind, we'd have a little bit more time with some of this stuff. That's the painful part.

PATRICK FARNSWORTH: I recently interviewed Will Falk, an environmental activist and a lawyer, about his time visiting the Unist'ot'en Camp in the Wet'suwet'en Nation, and we talked about his experience of being on that land. Then I also interviewed Dr. Karla Tait, program director of the Indigenous Healing Center at the Unist'ot'en Camp and a member of that nation. I suspect that Will Falk, Karla Tait, and all these other people living on that territory are fighting the pipeline project in that region, are fully aware of what's happening on the planet. Yet they still see the sacredness of what they've been able to protect. They see the value of the nonhuman life that exists in that territory.

They're willing to put their bodies in the way of that pipeline, and they have built structures in the way. They also built the Indigenous Healing Center, trying to undo generations of colonialism in Canada. That work is so beautiful.

We can look to Indigenous cultures, not to idealize them, but to look to their cultures and others that have healthier ways of viewing their relationship with the natural world. Even in the face of all this, they're acting with dignity and maturity. And that, to me is what makes me so viscerally angry when I see shit like this pipeline.

What did you learn in your travels of various cultures and societies are grappling with this information?

DAHR JAMAIL: That's actually very well put, Patrick. One of the people I met while working on this book—he's in my last chapter—is Stan Rushworth. He's a teacher, an author, a healer, and an elder of Cherokee descent. He pointed out to me that in the West we have this idea of human rights: what are my rights? In Indigenous cultures, they have moral obligations to future generations and to Earth. So as we grow up and mature and continue to evolve, I think we should deepen our commitment to those two moral obligations.

Something else that he talks about in my book is how we choose to comport ourselves during this time, even if everything is lost. If humans are going extinct and all the worst-case scenarios come true, that doesn't change the fact that I still want take care and be a steward of land where I live, of different species, and try to do the right thing in any way that I can.

Pretend that it's 30 years from now. Let's say Rob's child comes up to me and says, "you were on the planet in 2019 and there was still Arctic sea ice, glaciers in the lower 48 states, and the Amazon was still largely intact and all of it is now gone. You knew all this was coming. What did you do in 2019?" I want to be able to say I did everything that I could. Part of that is selfish, so I can cover my own ass. But then part of it is selfless: I would put my body down if it was going to save a couple of other species. I want to live my life in a way that shows my priority is living in right relation to the earth.

This book was a profound experience. It brought people like Stan into my life to give me that information literally as I was writing the conclusion, because I had no idea how to conclude this book with the devastating science that's in it. I had no idea, and then the earth brought into my life this extremely wise, intelligent, articulate elder. He brought in a bunch of very important Indigenous stories, as well as his own words to help me in the book. This has essentially become a touchpoint, from which now I'm in this new place. I'm trying to live into this new way of being that was created by literally writing this book.

ROB SIEMETZ: You know, I don't really have many in-person conversations about things like this. It is refreshing to talk about this, and what I think about is that I'm gonna be a father now any day. That's the ultimate challenge is to me, to my son: what am I going to do for his generation and for him? I appreciate everything that you're doing with the book and the journey that you're chronicling. At the end of the talk last night, you posed a challenge to everybody: where's your spot to connect? That's something I'm going to be thinking about a lot in the next couple days.

DAHR JAMAIL: I think that we're all morally obliged. Again, it keeps coming back to what I keep referring to as a moral obligation to the younger and future generations. I'm in a position now where I can help, and I feel like the onus is on us to protect them, to educate them, to prepare them and help them for this extremely challenging time that the world is in now. I see that as another layer of this moral obligation, because when I talk about being morally obliged to be a steward and help work for the earth, that includes all beings, including humans.

Like Dahr and Rob, I'm constantly trying to work this out of my mind and heart, and there are no easy answers on how to proceed, certainly not any that are universal in scope. Each of us must come to this in our own way. Grieving what is lost, treasuring and fighting for what's left, and accepting what is unfolding are part of our sacred responsibilities.

Conducting these interviews and putting this book together has been a means for me to explore some of the possible answers to how and why we arrived at this sorrowful place. I can only hope with this understanding we can then rise to the occasion with dignity, love, solidarity, and righteous power, for the living Earth and for our fellow human beings. This takes on countless forms. To discern what those acts may be, and to know how to act accordingly in these times of trouble, will demand the best from us. This is no time to look away, we have looked away for too long already.

THE TIMES ARE URGENT, LET'S SLOW DOWN.

WE LIVE AND THRIVE IN THE ORBIT OF BEINGS THAT ARE GREATER THAN US. AND IF WE HAVE TO MEET A WORLD, IF WE HAVE TO ACTUALLY COME TO TERMS WITH A WORLD THAT IS GREATER THAN US, THEN WE HAVE TO STOP THINKING IN TERMS OF ANTHROPOCENTRICITY, OF HUMAN-CENTEREDNESS.

BAYO AKOMOLAFE

There is only one way to end this book, and that is with lecturer, (post)-activist, and author Bayo Akomolafe. In this last segment, Bayo elaborates on what it means to "slow down" in a time of accelerating change as we run headlong into the "human epoch," the Anthropocene.

As Bayo so beautifully expressed in the essay "Coming Into the Sanctuary":

> I want to tell them about a vaster sanctuary I am coming to know: not one where you walk in broken and leave mended, but one where you walk in mended and leave broken. Where you walk in eloquent and leave with a lisp. Where the cavorting gait of the confident is disciplined by the hand of god upon the thigh—dislodging the bones. I want to tell them that in these times of painful losses, of starving bellies and emaciated hopes, where the lingering spectacle of destruction shadows the lands and the story of our godhood still gnaws at the bones of the world, we cannot afford those knowledges that presume our superiority. We cannot practice escape any longer—if we are to survive. We cannot cleanse ourselves of our sins or hope for the parting of the clouds to bring a convenient saviour. I want to tell them we must now gesture toward hopes and worlds beyond modern imaginaries. Beyond humans. Beyond the intelligible. Beyond our usual ways of making sense. I want to tell them that we must go to the edges, toward the hedgerows teeming with hagodays and gargoyles and stuttered beings, and learn to witness with-ness the world we once banished to the peripheries of significance.

PATRICK FARNSWORTH: I think it was about a year ago when I first contacted you and I think you had just released your book, *These Wilds Beyond Our Fences*. I wanted to continue our conversation because I enjoyed our first conversation so much. You use the term Anthropocene to describe this era that we've entered into, this age of human-centeredness. I talk about colonization, decolonization, capitalism, and radical political theory—I get into all these different territories, and it is overwhelming to view it through that lens all the time. I feel like intellectually and emotionally I'm up against something I can't get my mind around. It still is hard for me to really comprehend all the things that are happening right now.

BAYO AKOMOLAFE: Right.

PATRICK FARNSWORTH: It's almost too much. And I had this thought—I need to talk to Bayo again.

BAYO AKOMOLAFE: [laughing].

PATRICK FARNSWORTH: You have a way of grounding me, and I think you do that with so many other people as well. You tend to ground people in what's valuable and important, reminding people that the times are urgent, but we need to slow down.

BAYO AKOMOLAFE: Yeah.

PATRICK FARNSWORTH: I think many people, especially those that listen to my podcast, are feeling emotionally, intellectually, and maybe physically overwhelmed as well. How do you approach the subject when somebody like me asks you how we can understand this moment we're in? Where would you begin?

BAYO AKOMOLAFE: I'm usually very hesitant to create an easy convenient narrative, like "this is the age of..." and therefore why we experience this. Or this is "the age of..." and that's the reason why we're not experiencing certain things at the moment. Sometimes it's disingenuous and just overreaching, but at this point I actually feel that there are certain dynamics in our globalizing civilization that create the feeling, the sensation, this grounded embodied feeling of overwhelm. Maybe that's because we are collectively—and I mean a nuanced we— experiencing and coming to the end of the project of modern man, the indexing man, the one who wields the weapon of language and tries to reduce and encode all of reality in one fell swoop. And it seems ironically in responding to your question we are trying to do it again.

PATRICK FARNSWORTH: [laughs].

BAYO AKOMOLAFE: Part of the DNA of modernity is basically that, if we give it a shot, we can actually understand everything. There is no place that is sacrosanct. There is no place that is too sacred that we cannot go to. There is no mountain we cannot level. There is no tree we cannot remove to plant our highways, and the universe is just this space waiting for us to discover it.

We try to index everything, to finally categorize it, to say "we've got this one and we have this one in the bag." This manifests in the architecture of our society, in the ways that we frame work, in the ways we think about education, in the ways we think about citizenship, and even ironically in the way we think about justice. It's representationalism. It's this mirror image idea, that if we are pure on the inside, the world will straighten itself on the outside. "Be who you want to be and the world would follow suit."

This binary, this dualism, enforces and commends to us righteousness: we should be good people, and if we were good people we wouldn't be suffering from climate change. If we were good people, maybe black people won't be shot in the street. If we were good people, none of this would be happening. There's this burden of righteousness.

There's this drive to index further and further, and it seems like we're not facing up to the inhospitality of the world. The wish is that the world is hospitable, but it is also in many senses inhospitable. It will not bend all the time, it will not give way all the time. It will resist, it will push back. But we as blind modern citizens, gestating

I LIKE THE WORD HUMBLE AND I LOVE THE WORD HUMILIATED. I LIKE HOW THEY BOTH SPEAK ABOUT HUMANS, OF EARTH, OF FALLING TO THE GROUND.

in modernity, we are continually pushing, continually looking for the Golden Fleece or other mythological artifacts. We're looking for a way to finally arrive. And even I, speaking with you now, am also on that ironic journey, trying to figure things out, trying to balance things. We tend to find new ways of expressing this ontological impossibility of balancing, of arriving, of noticing the world differently, of being good people, or of finally becoming true. All of this defeats us and overwhelms us.

So what I like to say to people is (apart from the shrug of the shoulder, which is rich philosophy in my opinion): we're being defeated. We live and thrive in the orbit of beings that are greater than us. And if we have to meet a world, if we have to actually come to terms with a world that is greater than us, then we have to stop thinking in terms of anthropocentricity, of human-centeredness.

We have to learn to compose ourselves. This will not make us righteous. This will not address those political hot button issues like representation or the #MeToo movement. It will not always satisfy everyone. We will be offended most of the time. But here is the deal: the world is incredibly messy. That is not a justification for not doing all we can. But it is maybe a shocking reminder that we are not going to finally get this together. We are limited.

PATRICK FARNSWORTH: You've described it as needing to find our right place within this bigger thing, and that this global phenomenon is a symptom of a deeper problem, a deeper disconnection that we have. In a talk of yours, you were describing speaking to an elder in Nigeria. And this person said what's happening is because we felt like we could build roads everywhere. That we could expand our abilities and our reach, and the planet is responding to that.

My fear, and I think the fear a lot of people have as well, is that human beings may not ultimately be a part of the picture in the future. I think that's what scares people the most. When you talk about being a part of the compost heap, that to me is such a wonderful way to describe it. Sometimes the things that are part of the compost heap create something that looks totally different from anything that came before it. I think that what scares people about facing these crises is that we can't solve this. This isn't something that is meant to be solved, like every other great problem that humanity has tried to face. That's not really what it's asking of us at all.

I sense there's a real awakening, a lot of people coming into this place you're talking about. But even in the face of that, the fear is that there's so many others that aren't going to be there with us when this unfolds. Maybe I'm speaking more of my own fear. Maybe I'm projecting my own fears onto others right now. But I guess I'm sort of treating you a bit like a therapist right now, and I'm sorry about that. I figured this would happen at some point. I would start saying stuff like this.

BAYO AKOMOLAFE: [laughing}. So what I hear in you speaking about figuring things out, I guess it's a very difficult picture: a messiah piercing through the clouds, standing majestically, and only a few people know he's around. And some gather to him. And he says, get the word out. Just let everyone know I'm back. But they don't have much time to do it. And not everyone will know. This is something that disturbs me about this picture of arrival and figuring things out, because it seems to take for granted the idea that there is a single lyric. If I speak radically, poetically here, it assumes there is a standard to rise to, that the universe could be played on a single note, and any other thing that is away from that note is a deviation from what ought to be.

When we speak about figuring things out, we're articulating a product of starvation that demands evangelical outreach, presuming some exclusive-ism that we have the right to take on things that we don't. What that conveniently leaves out is context, and how people are embedded, entwined, and entangled with happenings that are larger than their supposed individuality. The modern gaze invites us to notice individuals trying to figure things out: they don't get it, or this guy is racist, or this guy is a climate denier. This is no justification for those things; I'm certainly of an ethics that subscribes to those points I have just enunciated. But the context is missing.

If we lose sight of the fact that we are part of assemblages, we are part of larger power forces that are creating a world, that are terraforming the planet, then we might be a bit more humble about figuring out stuff.

Let me see if I can be clearer. These things are hard to put into words, and sometimes are better left wordless. There is a story by Ursula K. Le Guin, a crazy, wonderful, beautiful story. It's called "The Ones Who Walk Away from Omelas." Ursula K. Le Guin describes the perfect city, basically heaven on earth—everything we could wish for, everything we fight for today.

Representation, access, good education and happiness, not the whimsical fleeting high, but a grounded hum of a stream of happiness that connects everything. Maybe I should say joy, but that seems too powerful a word.

People are acculturated. There's no exclusion. There's nothing of that sort. And Ursula directs our attention to an underground place in the city. She describes it as a dank, wet place. This kind of wardrobe place, you throw things into that you're not using. And in that wardrobe, in one of those wardrobes, in a cold, dark cellar underground in that same city, that wondrous happy city is a tiny "it." And I say "it" because Ursula refuses to describe the child as a girl or boy, because the child is so dehumanized, so emaciated, so battered and abused that it's no longer human, it's an it.

But the child can talk, and I'll just say she can talk because it's easier for me. She constantly pushes her hand outside of the small opening in the wardrobe and says, "please let me out. I'll be good. I'll be good. Just please let me out." And once in a while, the door opens to that room and someone slips in this horrible food on a plate, hastily put together, and closes the door again. Now, you might imagine reading such a story that obviously a morally enlightened city like this one doesn't have a clue about this creature, because if they did, they would rush to her aid. However, the reverse is the case: everyone knows. In fact, it's part of the education of the city to show everyone this child. So once a child does come of age, the parents take their children and they say, "look, that's the child."

And some can't take it. And they leave. And some just have to hold it, and they take it in there, because here is a riddle. The moment you try to show an atom of compassion to reach out to the child, the moment you try to alleviate the suffering, the moment you try to say it's going to be all right—the moment that happens, all the achievements, all of the goodness, all the wealth and happiness and success and cultural sophistication of this entire city would vanish in a heartbeat.

And so it's an ontological riddle, I'm not sure how that situation came to be. Ursula doesn't give us a clue. But that's the situation. Immediately, you try to save the child everything goes away. And facing that impossible situation, some people just walk across the field maddened by it and leave Omelas. Hence the title, "The Ones Who Walk Away from Omelas."

Now, the thing here is that Urusla doesn't tell the story in a tone of judgment. She doesn't say those that remain are evil and those that left are the good ones. I certainly would like to say this about the story: I feel that there is no figuring out, there is not another act of transgression, not another act of violence, that creates new shadows. Just to be, just to be present, just to be alive casts shadows upon something else.

So the very notion of figuring things out, looking for the ontological debt, assumes there's nothing beneath this, no shadows created by our being. This is it. And anyone who is not in this particular stream is missing or inadequate or wicked or sinful. And so instead we should be learning to be humble.

I like the word humble and I love the word humiliated. I like how they both speak about humans, of earth, of falling to the ground. Our collective global history speaks about the story of the need for transcendence, the urge to rise above the dirt, the planet and to see ourselves as separable from the world around us, from the messiness and the tension and the contradictions and the inherent irony of being alive or being embodied.

If I could see that there is something central to the Anthropocene, it is the urgency of embodiment. It's the urgency of embodiment that we are in bodies, we are embodied, we are connected with geography, we are connected with the rocks. That means there is no figuring out. There is no escape. Climate change is just one aspect of this whole collective networking, and there is no shifting outside the mess in a clear way.

I LIKE TO SAY THAT WE ARE FALLING DOWN TO EARTH

AND WE WILL NOT
ARRIVE INTACT.

ABOUT THE INTERVIEWS

My hope is that the voices of the individuals featured in this text have been adequately represented, with respect to each of their perspectives and works. I must express deep gratitude to each of them for positively responding to my request to feature their interviews in this work. I admire each of them greatly.

BAYO AKOMOLAFE

Episode: #179 | We Will Not Arrive Intact: The Times Are Urgent, Let's Slow Down

Bayo Akomolafe is a husband and father, as well as an international speaker, poet and activist for a radical paradigm shift in consciousness and current ways of living. Bayo is globally recognized for his unconventional, counterintuitive, and indigenous take on global crisis, civic action and social change. He is the author of *These Wilds Beyond our Fences: Letters to my Daughter on Humanity's Search for Home* and is the Executive Director and Coordinating Curator for The Emergence Network.

http://bayoakomolafe.net

LIYAH BABAYAN

Episode: #131 | The Other: Genocide; Life After

Liyah Babayan is an entrepreneur, activist, Armenian refugee, and the owner of Ooh La La! consignment boutique in Twin Falls, Idaho. She is the author of *Liminal: A Refugee Memoir,* which recounts the profound, disturbing, and moving experiences of her childhood fleeing the pogroms enacted against the Armenian minority population in Baku, Azerbaijan, in the midst the collapse of the Soviet Union in 1989-1990.

https://liyahbabayan.com

PAUL BECKWITH

Episode: #191 | Climate Casino:
Humanity, Stand Up Before The Pain Is Too Great

Paul Beckwith is a physicist, engineer, and part-time professor at the University of Ottawa. He earned a Masters in Science in Laser Optics/Physics, and a Bachelors of Engineering in Engineering Physics. He is a known and respected creator of entertaining and comprehensible videos of sometimes daunting subjects, especially in climate system science, meteorology, oceanography and earth sciences on YouTube.

https://paulbeckwith.net

JASPER BERNES

Episode: #198 | Sacrifice Zones:
Between The Devil & The Green New Deal

Jasper Bernes is the Managing Editor of Commune Magazine and Founding Editor of Commune Editions. He is the author of *The Work of Art in the Age of Deindustrialization* and two books of poetry: *We Are Nothing and So Can You,* and *Starsdown.* He lives in Berkeley, California with his family.

https://jasperbernes.net

JOE BREWER

Episode: #193 | Invisible, Sacred Work:
The Management Of Planetary Collapse

Joe Brewer is a change strategist, complexity researcher, cognitive scientist, and evangelist for the field of culture design. He has a background in physics, math, philosophy, atmospheric science, complexity research, and cognitive linguistics. Awakened to the threat of human-induced climate disruption while pursuing a Ph.D. in atmospheric science, he switched fields and began to work with scholars in the behavioral and cognitive sciences with the hope creating large-scale behavior change at the level of global civilization.

https://medium.com/@joe_brewer

SHANE BURLEY

Episode: #181 | The Violent Myth Of White Erasure: Terror In Christchurch

Shane Burley is a writer and filmmaker who regularly reports on far-right movements in the US, as well as anti-fascist resistance, workers rights, and class struggle. He is the author of *Fascism Today: What It Is and How to End It* from AK Press, and his work has appeared in numerous publications, such as Jacobin, Truthout, and Commune Magazine.

https://www.shaneburley.org

RAMON ELANI

Episode: #185 | The Gods Have Fled:
The Home As A Site Of Defiance Against Modernity

Ramon Elani is an acausal, anti-modern, heathen poet and author. He holds a Ph.D in literature and philosophy. He lives with his family among mountains and rivers in Western New England. He follows the way of wyrd.

https://thetigersleap.wordpress.com

SILVIA FEDERICI

Episode: #106 | Caliban And The Witch: The Body In The Transition To Capitalism

Silvia Federici is an Italian-American scholar, teacher, and activist from the radical autonomist feminist Marxist tradition. She is the author of numerous books, including *Caliban and the Witch: Women, the Body, and Primitive Accumulation* and *Revolution at Point Zero: Housework, Reproduction, and Feminist Struggle*. She was co-founder of the International Feminist Collective, an organizer with the Wages for Housework Campaign, and was involved with the Midnight Notes Collective.

PETER GELDERLOOS

Episode: #166 | How Nonviolence Protects The State: An Analysis Of Early State Formation

Peter Gelderloos is an anarchist, activist, and the author of numerous books and essays relating to the subjects of anarchism and resistance movements, as well as historical analysis of early state formation in human societies. Some of his works include *What is Democracy?, How Nonviolence Protects the State, Anarchy Works, The Failure of Nonviolence: From the Arab Spring to Occupy,* and *Worshipping Power: An Anarchist View of Early State Formation*. Peter currently resides in Spain.

JOHN HALSTEAD

Episode: #148 | The Dying God:
Learning To Die In The Anthropocene

John Halstead is a pagan writer, (former) activist, and author of *Another End of the World is Possible*, in which he explores what it would really mean for our relationship with the natural world if we were to admit that we are doomed. John is a native of the southern Laurentian bioregion and lives in Northwest Indiana, near Chicago. He is a co-founder of 350 Indiana-Calumet, which (until recently) worked to organize resistance to the fossil fuel industry in the region.

https://anotherendoftheworld.org

TAD HARGRAVE

Episode #142 | Reckoning With Whiteness:
Privilege & The Wounds Of Identity

Tad Hargrave is a man with an eclectic background. Tad is a founding member of several comedy troupes, dedicated almost two years of his life learning his ancestral language (Scottish Gaelic) in Nova Scotia and Scotland, runs a marketing website "Marketing For Hippies," worked as the Executive Director of the Canadian branch of Youth for Environmental Sanity, and writes for the blog "Healing From Whiteness: Remembering the Larger Story of History." Tad lives in Edmonton in Alberta, Canada.

https://healingfromwhiteness.blogspot.com

GORD HILL

Episode: #172 | No Justice On Stolen Land: Uniting Indigenous Resistance & Anticapitalist Revolt

Gord Hill, who also goes under the pen name Zig Zag, is a member of the Kwakwaka'wakw nation along the Northwest Coast of Canada. He is an anarchist activist, author and illustrator of numerous books/graphic novels, including *500 Years of Indigenous Resistance, The 500 Years of Resistance Comic Book, The Anti-Capitalist Resistance Comic Book,* and *The Antifa Comic Book.*

https://warriorpublications.wordpress.com

DR. GERALD HORNE

Episode: #120 | This Is America: The Apocalypse Of Settler Colonialism

Dr. Gerald Horne holds the Moores Professorship of History and African American Studies at the University of Houston and is the author of more than thirty books and one hundred scholarly articles and reviews, including: *The Apocalypse of Settler Colonialism: The Roots of Slavery, White Supremacy, and Capitalism in 17th Century North America and the Caribbean*. His research has addressed issues of racism in a variety of relations involving labor, politics, civil rights, international relations and war. Dr. Horne received his Ph.D. in history from Columbia University and his J.D. from the University of California, Berkeley and his B.A. from Princeton University.

NICHOLAS HUMPHREY

Episode: #87 | Heartland Deluge: The Floods Of An Unfamiliar Earth w/ Nicholas Humphrey

Nicholas Humphrey is a meteorologist and geoscientist, with a focus on extreme weather events and their connection to our destabilizing climate. Nick's goal is to communicate in an interdisciplinary fashion the serious risks from climate tipping points, extreme weather events, and ecological collapse. He graduated with a BS in Interdisciplinary Studies with a focus in societal impacts of extreme weather from South Dakota State University in 2013, and earned a MS in Geoscience—Applied Meteorology from Mississippi State University in 2016. Nicholas is based in Lincoln, Nebraska.

https://www.patreon.com/MeteorologistNickHumphrey

DAHR JAMAIL

Episodes: #154 | Another End Of The World Is Possible: Part One
#171 | The End Of Ice: Bearing Witness In The Path Of Climate Disruption

Dahr Jamail is an award-winning journalist who regularly reported on climate disruption and environmental issues for the online publication Truthout. Prior to his work reporting on environmental issues, Dahr was one of only a few unembedded journalists to report extensively from Iraq during the U.S. led invasion in 2003. Dahr is the author of multiple books, including *The End Of Ice: Bearing Witness and Finding Meaning in the Path of Climate Disruption*, released January of 2019.

http://www.dahrjamail.net

STEPHEN JENKINSON

Episode: #134 | Elderhood: Coming Of Age In Troubled Times

Stephen Jenkinson is a culture activist, teacher, author and ceremonialist. He is the author of numerous books, including *Come of Age: The Case for Elderhood in a Time of Trouble* and *Die Wise: A Manifesto for Sanity and Soul.* Stephen teaches internationally and is the creator and principal instructor of the Orphan Wisdom School, founded in 2010. With Masters degrees from Harvard University (Theology) and the University of Toronto (Social Work), he is revolutionizing grief and dying in North America.

https://orphanwisdom.com

DEZERAY LYN

Episode: #161 | Solidarity Not Charity: Mutual Aid Disaster Relief; A Factor Of Evolution

Dezeray Lyn is an activist, street medic, and co-organizer of Mutual Aid Disaster Relief (MADR), a "national network made up of many eco-activists, social justice activists, global justice activists, permaculturalists, community organizers, and others who are actively organizing around supporting disaster survivors in a spirit of mutual aid and solidarity." MADR is a decentralized network, defined by the character and creativity of a multitude of communities and drawn together by the collective commitment to stand in solidarity with those impacted by disasters and turn the tide in favor of climate justice.

https://mutualaiddisasterrelief.org

SINIŠA MALEŠEVIĆ

Episode: #165 | The Rise Of Organized Violence: The Brutalities Of National Identity w/ Siniša Malešević

Siniša Malešević is a Full Professor and Chair of Sociology at the University College, Dublin, Ireland. He is an elected member of Royal Irish Academy and Academia Europaea (the European Academy) and an elected Associated Member of the Academy of Sciences and Arts of Bosnia and Herzegovina. He is the author of numerous books, including *Grounded Nationalisms*, *The Rise of Organised Brutality: A Historical Sociology of Violence*, *Nation-States and Nationalisms: Organisation, Ideology and Solidarity, The Sociology of War and Violence*, *Identity as Ideology*, and *The Sociology of Ethnicity*.

https://sinisa2malesevic.wordpress.com

IAIN MCGILCHRIST

Episode: #115 | The Making Of The Western World: The Divided Brain

Iain McGilchrist is a psychiatrist, lecturer, and author of multiple books, including *Against Criticism, The Master and his Emissary: The Divided Brain and the Making of the Western World.* He is committed to the idea that the mind and brain can be understood only by seeing them in the broadest possible context, that of the whole of our physical and spiritual existence, and of the wider human culture in which they arise, the culture which helps to mould, and in turn is moulded by, our minds and brains.

https://iainmcgilchrist.com

DR. WILLIAM REES

Episode: #125 | Marching Toward Collapse: Biophysical Limits & Our Cognitive Blindspots

Dr. William Rees is a human ecologist, ecological economist, Professor Emeritus and former Director of the University of British Columbia's School of Community and Regional Planning in Vancouver, Canada. Dr. Rees is the originator and co-developer of the Ecological Footprint Analysis, and the co-author (with Mathis Wackernagel) of *Our Ecological Footprint: Reducing Human Impact on the Earth*, an exploration of this concept. The Ecological Footprint concept has become the world's best-known metaphor for the human load (the resources required of ecosystems to maintain our current mode of living) on the planet. Dr. Rees has authored (or co-authored) more than 150 peer reviewed papers and book chapters, and numerous popular articles on humanity's (un)sustainability conundrum.

http://williamrees.org

JOHN H. RICHARDSON

Episode: #199 | Kaczynski Moments: A Generation Of Unabomber Acolytes (& Apostates)

John H. Richardson is a journalist and former Writer at Large for Esquire Magazine. He is the author of three books, including *My Father The Spy, An Investigative Memoir.* He was born in Washington D.C. in 1954, and grew up in Athens, Manila, Saigon, Washington, Seoul, Honolulu, and Los Angeles. John has worked at the Albuquerque Tribune, The Los Angeles Daily News, Premiere Magazine, New York Magazine, and Esquire Magazine and taught at Columbia University, the University of New Mexico, and Purchase College.

https://johnhrichardson.com

FRANCISCO SANCHEZ-BAYO

Episode: #194 | Insect Apocalypse: Drivers Of The Worldwide Decline Of Entomofauna

Francisco Sánchez-Bayo is an ecotoxicologist and terrestrial ecologist, with a focus on risk assessment of pesticide contaminants on organisms, particularly on birds and aquatic ecosystems, and fate and transport of neonicotinoids and other agricultural chemicals in the environment. He is a former researcher at the University of Sydney, and currently works for the Department of Environment and Energy of Australia.

http://sydney.academia.edu/FranciscoSánchezbayo

ROB SEIMETZ

Episodes: #171 | The End Of Ice: Bearing Witness In The Path Of Climate Disruption
#196 | We're Doomed!: Of Our Elaborate Plans, The End

Rob Seimetz is a writer and former host of Moving Forward on the Progressive Radio Network. He currently lives Portland, Oregon in the United States.

MICHAEL SLIWA

Episode: #196 | We're Doomed!: Of Our Elaborate Plans, The End

Michael Sliwa is the author of *Chasing A Different Carrot: A Manifesto For The Predicament Of Privilege* and is one of the foremost speakers on simple living. He and his wife Karen are former educators who left behind their careers and most of their worldly possessions in order to pursue a life of genuine connectivity. Michael is the Co-Founder and Development Specialist for TRUality, and is the former co-host of Nature Bats Last on the Progressive Radio Network.

https://michaelsliwa.wordpress.com

DR. KARLA TAIT

Episode: #169 | Heal The Land, Heal The People: The Unist'ot'en Healing Center

Dr. Karla Tait is a clinical psychologist, and is presently employed as the Mental Wellness Advisor for the Northern Region with the First Nations Health Authority, serving the 54 First Nations in Northern BC. Dr. Tait is the Director of Programming for the Unist'ot'en Healing Centre, a collective effort by her Hereditary House Group to revitalize culturally-based, wellness and holistic healing practices that are centered on connection to the land. Dr. Tait belongs to the Gilseyhu clan (Big Frog) clan of Yahstowilcus (the Dark House). Her Dini Zi' (House Chief) is Unist'ot'en, Knedebeas, Warner William, maternal grandmother is Weli', Catherine Michell, maternal grandfather was Tsayu, Wigitemstochol (Dan Michell), mother is Brenda Michell-Joseph, father is Rene Tait and stepfather is Melvin Joseph. Her father clan is Laksilhyu (small frog/cariboo).

http://unistoten.camp

HOLLY TRUHLAR

Episode: #186 | Mainstream Psychology Can Go Fuck Itself: Managing Collapse Awareness

Holly Truhlar is a lawyer, grief therapist, and community builder. She's integrated her doctorate in law with a masters in transpersonal counseling psychology to create a model for personal and collective liberation. This model, called Radical Attunement Work (RAW), moves people and groups from dominating and exploitative ways of being into attuned and transformational relationships. Her work lies at the convergence of ancestral grief, collective trauma, conscious collapse, and social justice.

https://www.hollytruhlar.com

RHYD WILDERMUTH

Episode: #197 | All That Is Sacred Is Profaned: Marxism, Paganism, & History As Process

Rhyd Wildermuth is a co-founder and the publishing manager of Gods & Radicals Press. He's also a bard, poet, and theorist. He has written numerous essays on paganism, anti-capitalist political theory and analysis, and is the author of *Witches In A Crumbling Empire*, and most recently *All That Is Sacred Is Profaned: A Pagan Guide To Marxism.* He is the co-host of Empires Crumble, a podcast on politics, culture, history, and magic with Alley Valkyrie. After spending much of his life within the borders of the United States, he now resides in Luxembourg.

https://paganarch.com

XABAT / MAKE ROJAVA GREEN AGAIN

Episode: #189 | Make Rojava Green Again: The Ecological Society w/ The Internationalist Commune Of Rojava

Xabat is a representative of the Internationalist Commune of Rojava and the Make Rojava Green Again campaign in cooperation with the Ecology Committee of the Cizire Canton in Northeastern Syria.

https://makerojavagreenagain.org

ABOUT THE AUTHOR

PATRICK FARNSWORTH

Patrick Farnsworth is a long-form interviewer and host of Last Born In The Wilderness, a weekly-released podcast that covers such broad topics as anthropogenic climate change, radical political theory and praxis, animism, psychedelics, and current events.

https://www.lastborninthewilderness.com

ABOUT GODS&RADICALS PRESS

Gods&Radicals Press is a not for profit pagan anticapitalist publisher founded in 2015. For our full catalogue of works, please visit our website and online journal:

abeautifulresistance.org

LAST
BORN
IN THE
WILDERNESS

CPSIA information can be obtained
at www.ICGtesting.com
Printed in the USA
FSHW021702240720
71608FS